Galaxy S8 Manual for Beginners

The Recommended Galaxy S8 Guide for Seniors, Beginners, & First-time Smartphone Users

Joe Malacina

Galaxy S8 Manual for Beginners

Copyright 2017 by No Limit Enterprises, Inc., All Rights Reserved

Published by No Limit Enterprises, Inc., Chicago, IL

No part of this publication may be reproduced, copied, stored in a retrieval system, or transmitted in any form or by any means, electronic, digitally, mechanically, photocopy, scanning, or otherwise except as permitted under Section 107 or 108 of the 1976 United States Copyright Act, without the express written permission of No Limit Enterprises, Inc., or the author. No patent liability is assumed with respect to the use of information contained within this publication.

Limit of Liability & Disclaimer of Warranty: While the author and No Limit Enterprises, Inc. have used their best efforts in preparing this book, they make no representations or warranties with respect to the accuracy or completeness of the contents of this book and specifically disclaim any implied warranties of merchantability or fitness for a particular purpose. No warranty may be created or extended by sales representatives or written sales materials. The advice, strategies, and recommendations contained in this publication may not be suitable for your specific situation, and you should seek consul with a professional when appropriate. Neither No Limit Enterprises, Inc. nor the author shall be liable for any loss of profit or any other commercial damages including but not limited to: consequential, incidental, special, or other damages.

The views and/or opinions expressed in this book are not necessarily shared by No Limit Enterprises, Inc. Furthermore, this publication often cites websites within the text that are recommended by the author. No Limit Enterprises, Inc. does not necessarily endorse the information any of these websites may provide, or does the author.

Some content that appears in print may not be available in electronic books and vice versa.

Trademarks & Acknowledgements

Every effort has been made by the publisher and the author to show known trademarks as capitalized in this text. No Limit Enterprises, Inc. cannot attest to the accuracy of this information, and use of a term in this text should not be regarded as affecting the validity of any trademark or copyrighted term. Galaxy S8® is a registered trademark of Samsung Group. Android is a registered trademark of Google (Alphabet Inc.). All other trademarks are the property of their respective owners. The publisher and the author are not associated with any product or vendor mentioned in this book. *Galaxy S8 Manual for Beginners* is an independent publication and has not been authorized, sponsored, or otherwise approved by Samsung Group or Alphabet Inc. Other registered trademarks of Samsung Group and Alphabet Inc. referenced in this book include: Bixby, Google Assistant, and Chrome.

Warning & Disclaimer

Every effort has been made to ensure this text is as accurate and complete as possible. This text does not cover every single aspect, nor was it intended to do so. Instead, the text is meant to be a building block for successful learning of the subject matter of this text. No warranty whatsoever is implied. The author and No Limit Enterprises, Inc. shall have neither liability nor responsibility to any person or entity with respect to any losses or damages arising from the information contained within this text.

Contact the Publisher

To contact No Limit Enterprises, Inc. or the author for sales, marketing material, or any commercial purpose, please visit www.nolimitcorp.com, or email us at info@nolimitcorp.com.

Galaxy S8 Manual for Beginners

Publisher: No Limit Enterprises, Inc.

Author: Joe Malacina

Cover art by: Erik Christian Voigt

ISBN: 978-0-9989196-5-2

Printed in the U.S.A.

Table of Contents

Introduction 9

Chapter 1 – About the Galaxy S8 11
Key Terms 11
Different Galaxy S8s and Wireless Carriers 12

Chapter 2 – Galaxy S8 Layout 13
Inserting a SD or SIM Card 14
Charging the Galaxy S8 14
Turning the Galaxy on or off 14

Chapter 3 – Getting Started 17
First-Time Setup 17
Chapter 4 – Navigation 23
Apps Page 24
Bixby Screen 24
Notification Bar 24
The 3 Main Buttons (Navigation Bar) 25
The Keyboard 28
Connecting to Wi-Fi 29

Chapter 5 – Google and Samsung Account 31
Google Account 31
Samsung Account 31

Chapter 6 – Contact List 33
Importing Contacts 33
Creating Contacts 33
Browsing Contacts 35
Contact Groups 35
Managing Contacts 37

Chapter 7 – Phone Calls 39
Making a Call 39
Receiving a Call 42
Call Functions 43
Recent Calls 44
Voicemail 45

Chapter 8 – Text Messaging 47

Messages App 47

Sending a Text Message 48

Receiving a Text Message 49

Group Conversations 49

Attachments 50

Managing Messages 50

Chapter 9 – Email 53

Adding an Email Address to your Galaxy 53

Checking your Email 55

Viewing Email 57

Sending Email 58

Managing Email 59

Chapter 10 – Web Browsing 61

Visiting Web Pages 61

Using Chrome Tabs 65

Chapter 11 – Using Your Camera 67

Camera Modes 68

Filters & Effects 69

Taking a Photo 69

Recording a Video 70

Chapter 12 – Photos & Videos 71

Gallery App Overview 71

Samsung Cloud 72

Browsing Photos 72

Photo Albums 72

Sharing Photos 73

Managing Photos 74

Editing a Photo 75

Chapter 13 – Galaxy Security 83

Settings 83

Other Security Settings 88

Chapter 14 – Personal Settings 89
    Setting your Wallpapers 89
    Themes & Other Visual Customizations 92
    Sounds & Ringtones 93

Chapter 15 – Apps 95
    The Play Store 95
    Browsing the Play Store 95
    Browsing by Top Charts 97
    Browsing by Category 98
    Searching for an App 99
    Downloading Apps 99
    Important Information about Downloaded Apps 100
    Downloading Media Content 100
    Additional Resources for Play Store Apps & Media Content 102

Chapter 16 – The Home Screen 103
    Apps 103
    Widgets 103
    Organizing Home Screens, Apps, & Widgets 104

Chapter 17 – Notifications 115
    Overview 115
    The Notification Bar 116
    Examples of Notifications 117
    Managing Notifications 117

Chapter 18 – The Notification Bar 119
    Quick Controls 119

Chapter 19 – Bixby & Google Assistant 123
    Setting up Bixby 123
    Using Bixby 123
    Setting up Google Assistant 124
    Using Google Assistant 125
    Bixby vs. Google Assistant 126

Chapter 20 – Native Apps 127
    Different Apps with Different Carriers 127

Play Music 127

Google Maps 129

Internet 131

My Files 132

Clock 133

Calendar 134

Chapter 21 – Tips & Tricks 137

    Backing up your Device 137

    Taking a Screenshot 138

    Background Apps 138

    Multi Window 139

    Restart your Galaxy 140

    Copy and Paste 140

    Status Bar 141

    Updating your Galaxy 142

Chapter 22 – More Resources 143

Conclusion 145

Appendix A – Recommended Apps 147

Appendix B – List of Common Functions 149

Introduction

Congratulations! So you have decided to take the first step, in fact the only step, needed to learn how to use your Samsung Galaxy S8 smartphone. Perhaps you do not even have the Galaxy S8 yet, and just want to see how it works before you decide whether to get one. Either way, this book will teach you everything you need to know on operating the Galaxy S8. I want to take this time to thank you for reading the introduction as it contains important information as to how this book is structured, and how you should read it. You see, this book was written for the perspective of a complete beginner to modern smartphones. In other words, if you have never used a smartphone in your life, that will be no detriment when reading this book. I understand that many people are still buying their first smartphone in the year 2017, and need to be shown from the ground up the basics of operating their device. That is exactly what this book sets out to accomplish. I teach you not only how to do specific functions on your Galaxy S8, but I will teach you the building blocks of using any *smart* device. Consequently, when you are finished reading this book you should be able to pick up any modern smartphone or tablet and have a general understanding of how it works and how to accomplish tasks on it.

What about the seasoned smartphone users? This book is still good for you too. Besides the basics, this book delves into intermediate and advanced functions of the Galaxy S8, particularly in the later chapters. It also covers new functions that previous Galaxy smartphones could not do, and explores the Nougat operating system thoroughly.

This book is structured so that basic concepts covered in beginning chapters will be used in later chapters. Therefore, I highly recommend that beginners read the first few chapters, instead of skipping ahead to exactly what you want to learn. You may miss out on essential tidbits of information that I will not cover in detail in later chapters.

Lastly, it is important to know that this book is not an all-encompassing instruction manual to the Galaxy S8. If that were the case the text would be over 300 pages long and chances are you would never read it. Instead, this book covers all the basics as well as the most important and common functions, and you can read it in just a few hours. Everything is explained in a step-by-step approach with detailed illustrations and prudent background information to provide clarity and build your technological intuition. When you are done reading this book, you will not need to memorize the steps to perform a function; you will have the intuition and knowledge to figure it out quickly. That is the core of what this book sets out to accomplish.

With all that said, let us begin.

Chapter 1 – About the Galaxy S8

So now we are ready to learn about the Samsung Galaxy S8. Before we get in to the actual instruction portion of this book, let's examine some background information and key terms regarding the Galaxy S8. First, the Samsung Galaxy S8 phone is classified as a smartphone, which means that it can generally do everything a cell phone can do plus some additional features. With your Galaxy S8, you can make and receive calls, send text messages, check your email, and use apps. You can do all this with a phone that utilizes a touch screen. Now let us familiarize ourselves with some key terms that are important to remember. I will be referencing these terms often throughout the book.

Key Terms

Google Account – This term is probably the most important throughout the entire book, and also one of the most confusing. Your Google account is an account you will need to setup on your Galaxy S8 in order to use the phone to its full extent. You will need a Google account in order to download apps, backup your device, and sync contacts. Creating a Google account is fairly simple, and I will cover setting one up in Chapter 3.

Samsung Account – A Samsung account is another type of account I am going to highly suggest you sign up for. Creating a Samsung account allows you to use certain features on your S8 including Find My Mobile and Samsung Hub. I will cover how to setup a Samsung account in Chapter 3.

Apps – Apps are programs on your Galaxy S8 that can perform tasks. Your Galaxy S8 comes with many apps pre-installed, and your cellular provider sometimes installs several apps of their own on your device. You can find and download additional apps in the Play Store, which I will cover later in this book. Nearly every aspect of the Galaxy S8 is part of an app, including making phone calls and sending text messages. Apps appear as ovals on your Galaxy's home screen and Apps page (Figure 1.1).

Home Screen – Throughout this book, you will see the term home screen used often. Your home screen is the just the main screen of your phone that shows your apps and widgets. Your phone has multiple home screens that you can switch between by swiping left and right.

Android – The term Android is the name of the operating system running on your phone. Your operating system is the software that controls how your phone operates. For the Galaxy S8, the default installed operating system is Android 7.X Nougat, where the X will be some number. For the purposes of this book, just know that your Galaxy S8 should be running the Nougat or newer operating system, which is any version of Android that starts with a 7 or higher number.

Chapter 1 | About the Galaxy S8

Figure 1.1 – The Home Screen

Different Galaxy S8s and Wireless Carriers

There are several different Galaxy S8s available on the market. The two most notable are the Galaxy S8 and the Galaxy S8+. The main difference between these two models is the size of the device and the size of the screen. Other than that, both devices will operate nearly exactly the same. Thus, this book will work for any Galaxy S8 model.

Furthermore, some wireless carriers install their own set of apps on the Galaxy S8. For instance, you may see multiple apps on your device that can send and receive text messages. Or, you may see multiple apps that can provide GPS and driving directions. Regardless of what these carriers install on your device, these apps are not necessary to perform basic functions. The default Galaxy S8 apps for text messaging and the like are perfectly suitable to perform the basic functions, and this book will cover these default apps.

Chapter 2 – Galaxy S8 Layout
===

Now we enter the *instruction manual* portion of this book, and we start with the Galaxy S8 layout. Shown in <u>Figure 2.1</u> is the layout of a Samsung Galaxy S8.

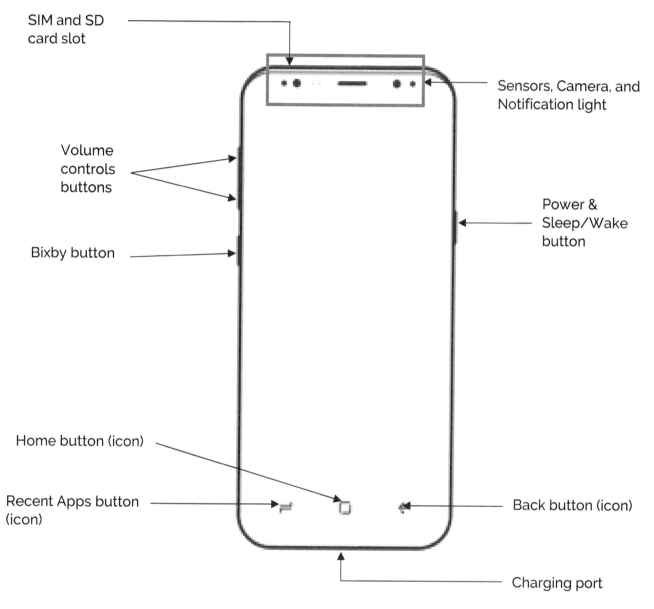

Figure 2.1 – Galaxy S8 Layout

On the right side of your Galaxy you can find the power button. Pressing and holding this button will turn the Galaxy on or off. While the Galaxy is on, pressing the power button will wake the Galaxy from the sleep slate or put the Galaxy into the sleep state. On the left side of the phone are two buttons. The large button near the top is the volume control button. Pressing the top half of this button will increase the volume while pressing the bottom half decreases volume. Below the volume control button is the

Bixby button, which activates the Bixby function of the Galaxy. Bixby is covered in Chapter 19 of this book.

At the top of the Galaxy on the front side are various sensors and lights. The largest sensor is the front-facing camera. Lastly, at the upper left is a small notification light that will blink when you have a notification.

At the bottom edge of the Galaxy is the charging port.

Finally, at the bottom of the front of the Galaxy according to Figure 2.1 is where you have your main navigation icons. Historically, on previous Galaxy phones these were actual buttons on the device. With the Galaxy S8, there are no longer buttons at the bottom. Instead, three icons have replaced the three main buttons, and these icons will appear on your Galaxy screen. These icons are the Recent Apps icon, Home icon, and Back icon. **Since these icons are essential to using the Galaxy S8 and are always available no matter you are doing with your device, I will refer to each icon as a button throughout the book. For instance, if I instruct to press the home button, that indicates to press the home icon.**

Inserting a SD or SIM Card

If you need to insert a SIM card or SD card into your Galaxy S8, you can use the slot at the top of the device. Simply push a paperclip end into the small circle and the tray will pop open. Insert the SIM card and SD card (if applicable) into the tray and slide back into place, until the tray locks. An SD card provides additional storage for your Galaxy while a SIM card grants it cellular service.

Charging the Galaxy S8

The Galaxy S8 can be easily charged by plugging the included charger into the bottom of the phone and connecting the other end into a standard USB port or charging dock. The Galaxy can be charged on any computer or device with a powered USB port. You can also use the charging port to plug your Galaxy into your computer.

Turning the Galaxy on or off

When the Galaxy is off, to turn it on simply press and hold the power button until the display screen lights up or the phone vibrates. Now wait several moments for your Galaxy to turn on completely.

At any time you can turn the Galaxy S8 off by pressing and holding the power button until Figure 2.2 appears. From here you can tap on Power Off to turn off the phone.

Chapter 2 | Galaxy S8 Layout

Tap to power off

Tap to restart Galaxy

Tap to enter Emergency mode

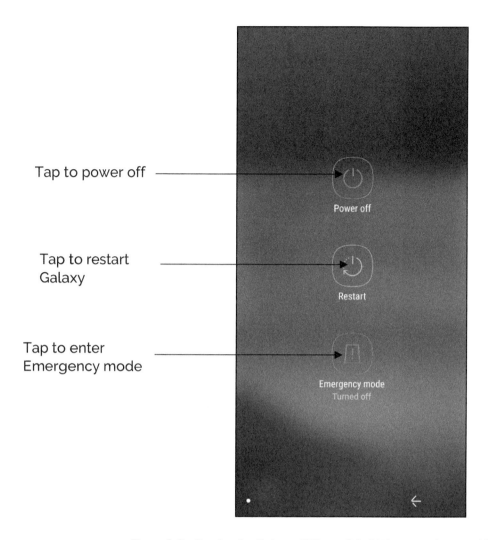

Figure 2.2 – Turning the Galaxy off (Press & hold the power button while Galaxy is on)

Chapter 3 | Getting Started

Chapter 3 – Getting Started

In this Chapter I will be covering the first-time setup procedure for the Galaxy S8. If you have already completed the first-time setup of your Galaxy S8, you can skip the first section of this chapter.

First-Time Setup

When you turn your Galaxy S8 on for the first time you will need to complete the first-time setup. This setup allows you to enter in your personal information, sign in to accounts, and personalize your settings. Here is the setup procedure:

1. Turn your Galaxy S8 on for the first time and wait for the setup to initialize. It may take several minutes.
2. The welcome screen will appear (Figure 3.1). Tap on START with your finger.

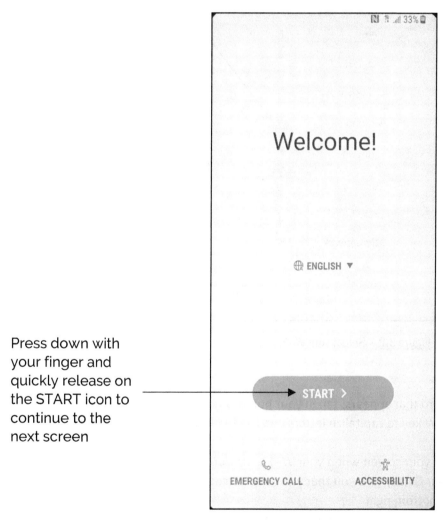

Press down with your finger and quickly release on the START icon to continue to the next screen

Figure 3.1 – The Welcome Screen (First-time Setup)

3. Next you will be prompted to select a Wi-Fi network in order to activate your phone. It is highly recommended that you setup your Galaxy at your home so you can connect to your home

wireless network. (If Wi-Fi is not available, you can tap on NEXT at the bottom to activate your phone using your cellular data network).
   a. Look for your home wireless network and tap on it with your finger (Figure 3.2).

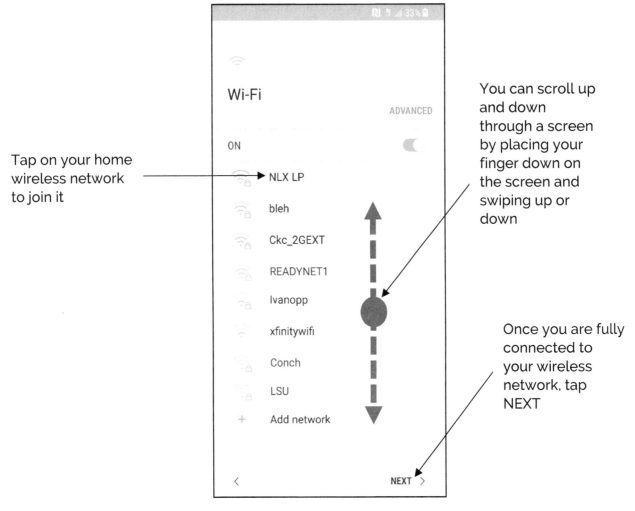

Figure 3.2 – Select your Wi-Fi network

   b. Using the keyboard that appears, tap in your home wireless network password. You can use the UPARROW key to capitalize letters and the !A# key to access symbols and punctuation.
   c. Tap <u>CONNECT</u> on your screen when your Wi-Fi password is entered.
   d. Tap <u>OK</u> when your Galaxy tells you that you have successfully connected to Wi-Fi.
   e. Tap <u>NEXT</u> at the bottom right.
4. The next screen will be the Terms and conditions screen. Here you can read through the terms and conditions at your leisure by tapping on the underlined word: <u>Terms and Conditions</u>. When you are ready to proceed, tap the <u>small bubbles</u> next to what you agree to, then tap <u>NEXT</u> at the bottom right (Figure 3.3).

Chapter 3 | Getting Started

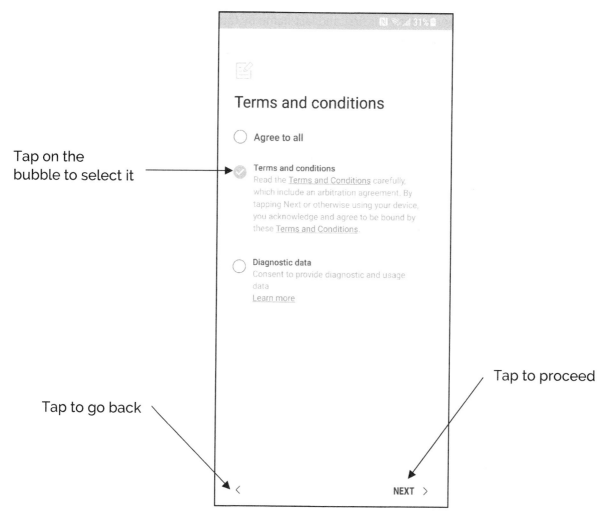

Figure 3.3 – Terms and Conditions

5. Next your Galaxy will quickly check for software updates. If an update is available, let your Galaxy update itself.
6. Next you will be asked to add your Google account. THIS IS A VERY IMPORTANT STEP, PLEASE FOLLOW THESE INSTRUCTIONS CAREFULLY. It is HIGHLY recommended that you sign in or create a Google account before proceeding. You can choose to skip this step if you wish, but doing so will cause you to miss out on a lot of features on your Galaxy S8.
7. If you know you already have a Google account, tap into the "email or phone" box and type in your email and phone number, followed by tapping on GO on your keyboard. Please note, if you have a Gmail email address, then you already have a Google account you can use. Just type in your Gmail email address, tap on GO, then type in your email password. If you do not have a Google account or are unsure, move on to step 7a. If you have successfully signed with your Google account, skip to step 8.
    a. If you currently do not have a Google account or are unsure, tap on "Or create a new account"

Chapter 3 | Getting Started

- b. On the next screen enter your first and last name by tapping into the corresponding boxes using your keyboard. Tap NEXT when your name is entered.
- c. Next enter your date of birth and gender. Tap into each corresponding box, then tap the corresponding option to select it. Please note, you can scroll through options by tapping on your screen and moving your finger up and down on the screen, as if flicking though a page. Tap NEXT at the bottom right when done.
- d. Next you will be asked to create a Gmail email address. The Gmail email address you create will be the email address that becomes your Google account. Use the keyboard and type in an email address you want to create. Tap NEXT at the bottom right when done.
- e. Next you will have to create a password for your Google account. Your password must contain at least 8 characters, and it is strongly advised that you use a mix of capitalization, letters, numbers, and symbols. THIS PASSWORD YOU CREATE IS VERY IMPORTANT. PLEASE WRITE IT DOWN SOMEWHERE SAFE. Tap NEXT when done.
- f. Next you will have to enter your phone number. Your current phone number may autofill. If it does not automatically fill in your number, enter it, then tap NEXT at the bottom right.
- g. Tap VERIFY
- h. Once your phone number is verified you will be brought to a Privacy and Terms screen. Read through this at your leisure and scroll all the way down to the bottom. Remember, to scroll simply place your finger down on the screen and then swipe your finger up or down continuously. When you have reached the bottom tap on I AGREE at the bottom right to agree to the terms.
- i. On the next page you will be asked to sign in. Tap NEXT at the bottom right.
8. The next screen will show you various Google services that you can enable for your account. I recommend enabling all available services by making sure each bubble is filled in. Then tap NEXT at the bottom right.
9. Now you should be completely signed in to your Google account.
10. The next screen will prompt you to setup protection for your device. There are many options to choose from and you can select whatever you wish. For now, tap on No, thank you and tap NEXT at the bottom right. (I will cover Security options later in this book)
11. Tap SKIP ANYWAY if prompted.
12. The next screen will ask you to review some additional apps. I recommend leaving this screen as it is, and then tap OK at the bottom right.
13. Next you will be prompted to setup a Samsung account. THIS STEP IS VERY IMPORTANT, PLEASE FOLLOW INSTRUCTIONS CAREFULLY. If you already have a Samsung account, tap on SIGN IN at the bottom, followed by signing in with your associated email address and password, then move to step 14. If you do not have a Samsung Account, I HIGHLY recommend creating one as you need one to use essential functions on your Galaxy. To create a Samsung account tap on CREATE ACCOUNT at the bottom and then move to step 13a.
    - a. Creating a Samsung account is very simple. You can use the same email you used to create your Google account. So if you just created a Gmail email address to create a Google account. Use that same Gmail address to create your Samsung account. This email may fill in automatically.

Chapter 3 | Getting Started

      b. Create a password for your Samsung account. You can use the same exact password you created for your Gmail address.
      c. Type in your First name, Last name, and ZIP code by tapping into each corresponding box.
      d. Tap the DATE OF BIRTH box and select your date of birth using the calendar. You can use the arrows to move between months.
      e. Tap the bubble next to Send me marketing information if you so choose. Tap NEXT.
      f. Read the Terms and Conditions and tap the corresponding bubble(s) to agree. Then tap AGREE at the bottom right. (I recommend that you agree to all)
      g. Next you will be asked to verify your phone number. If your phone number is not entered, enter it in. Tap on VERIFY. If verification fails, tap on SKIP.
      h. The next screen will welcome you to your Samsung Account. Tap on NEXT at the bottom right.

14. You will now be completely signed in to your Samsung account. Again, you may be asked to setup Phone Protection. If this appears, tap on No, thank you and then NEXT, for now.
15. The next and final screen will be the More useful features screen. I suggest leaving this screen as it is, and then tap FINISH at the bottom right.
16. If any other screens appear, just tap NEXT, CLOSE, or SKIP to move past them.

You have now completed the initial setup process of your Galaxy S8. You will now be brought to your home screen and you can finally start using your device!

Chapter 4 – Navigation

All you need to use your Galaxy are your fingers. Everything is based on the touch screen and the three main buttons on your Galaxy. Here at the home screen (Figure 4.1), which is just the screen that shows your apps and widgets, you can open an app, which is a program on your phone that can do tasks and enable features, by lightly touching down on the oval with your finger and releasing quickly. To then leave the app and return to the home screen, press down on the home button.

The home screen is actually made up of multiple screens that can be fully customized. To move through the different home screens, touch down on a home screen with one of your fingers, and then drag your finger to the left or the right, like you would be dragging a page, then release. If at any time you need to return back to the main home screen, just push down on the home button.

There will also be widgets on your home screen, which are just interactive functions of apps. For instance, the weather being shown on your home screen is a widget. You can tap on a widget to see more information.

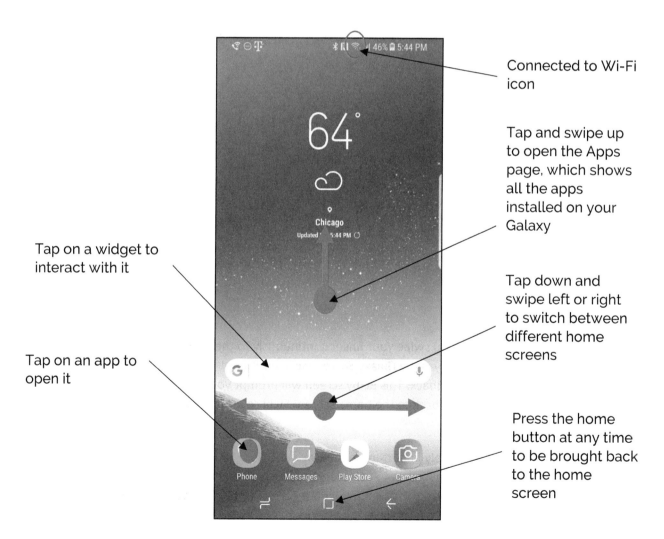

Figure 4.1 – The Home Screen

Chapter 4 | Navigation

Apps Page

To access the Apps page, which is a screen that shows all of the apps installed on your device, tap down on the home screen and swipe up. The Apps page is shown in Figure 4.2. You can return to your home screen by pressing the back button.

Figure 4.2 – The Apps Page

Bixby Screen

While on your home screen, if you swipe your finger continuously to the right you will eventually be brought to the Bixby screen. Bixby is the Galaxy S8's artificial intelligent assistant which can listen to your commands and provide feedback. This Bixby screen will prompt you to setup Bixby. For now, you can press the home button to go back to the home screen. I will cover Bixby later on in Chapter 19.

Notification Bar

The Notification Bar is an extremely important feature of the Galaxy S8. The Notification Bar can be accessed from any screen by tapping down at the very top of your screen and swiping down. Your Notification Bar will now be shown. This bar shows you recent notifications, allows you to alter some quick options, and gives you easy access to Settings. I will cover the Notification Bar fully in Chapter 18.

The 3 Main Buttons (Navigation Bar)

Now we return to discussing the three main buttons on your Galaxy S8 (Figure 4.3). These buttons can be accessed at any time while using your Galaxy, even when they are not shown on your screen. When on your home screen, you can see the three main buttons at the bottom of your screen.

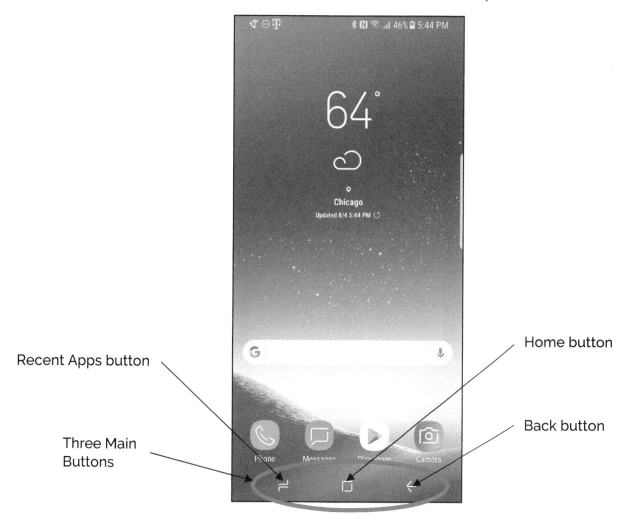

Figure 4.3 – The 3 Main Buttons on the Home Screen

The Recent Apps Button

The left button, is the Recent Apps button. Tapping on this will bring up a screen showing apps you have recently used (Figure 4.4). From here you can quickly go to a recently used app by tapping on its window. Alternatively, you can close an app completely by tapping and swiping it off the screen. To leave the Recent Apps screen, tap the back button.

Chapter 4 | Navigation

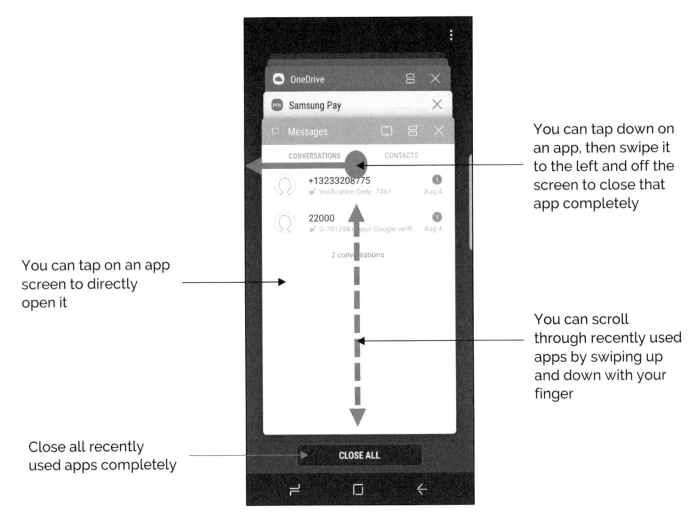

Figure 4.4 - Pressing the Recent Apps Button

The Home Button

The home button is the most important button/icon on your phone. Pressing the home button will always bring you back to your home screen when you are inside an app. The home button is shown as the square icon in Figure 4.3. Sometimes, you may not see the home button when using an app. You can still press the home button even if it does not appear. To do this, press on your home screen at the bottom where the home button is usually located. Do so with some additional force, and you will feel your phone vibrate slightly and the home button will be pressed. In fact, whenever you press the home button you will feel your Galaxy S8 vibrate slightly.

The Back Button

The back button is the button on the bottom right of your home screen (Figure 4.3). Pressing this button at any time will bring you back one action or screen. For instance, when you press the back button immediately after opening an app, you will immediately be brought back to your home screen. On the other hand, if you press the back button after browsing to a web page in Chrome, you will be brought to the previous page. We will be using the home button and back button extensively in this book.

Chapter 4 | Navigation

Accessing the Navigation Bar

As mentioned earlier, the home button is always available to press, even if it is not shown. The back and recent apps button however, do not work like this. In order to use these buttons when they do not appear, you will need to bring them up. Here is how you do it (Figures 4.5 & 4.6):

1. While using an app where the three main buttons are not shown, tap down at the very bottom of your screen and swipe up.
2. This will bring up the three main buttons, and you can now tap on any one you wish.
3. If you want this navigation bar to remain permanently, tap the small circle on the left hand side (RECOMMENDED).

Tap at the bottom of your screen with your finger or thumb, and swipe up

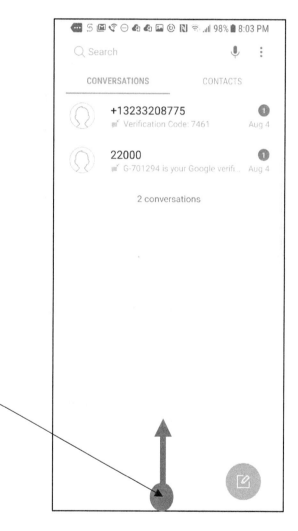

Figure 4.5 – Accessing the Navigation Bar (Step 1)

27

Chapter 4 | Navigation

Figure 4.6 – Accessing the Navigation Bar (Steps 2 & 3)

The Keyboard

The Galaxy S8 keyboard is a standard keyboard that offers predictive typing and some other features. Shown in Figure 4.7 is the general layout of the keyboard, which will automatically appear whenever you need to type.

Chapter 4 | Navigation

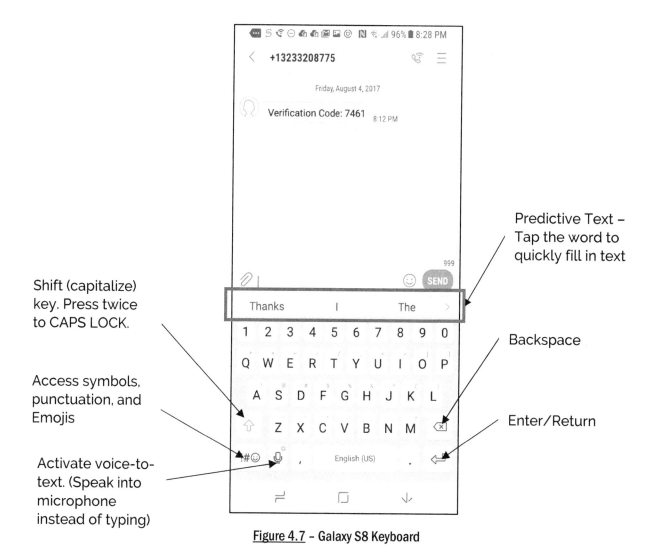

Figure 4.7 – Galaxy S8 Keyboard

When your keyboard appears, you can begin typing by just tapping the corresponding icons on the keyboard. Use the up arrow key to capitalize letters. You can also slide-type, which is dragging your finger to each letter while keeping your finger on the screen until the word is complete.

Connecting to Wi-Fi

You have probably already connected your Galaxy S8 to Wi-Fi when you first set up your device. Once you have connected your Galaxy to a certain Wi-Fi network you will never have to connect to it again, as your Galaxy will automatically connect to it whenever it is in range. I highly recommend connecting to a Wi-Fi network whenever you are able to, as Wi-Fi speeds are generally faster than cellular data and Wi-Fi does not count against your data limit on your cellular plan. Here is how you connect to Wi-Fi on the Galaxy S8:

1. From your home screen, tap down on a blank area of the screen and swipe up to bring up your Apps page.

2. Find the Settings app and tap it.
3. Tap the Connections tab.
4. Make sure the slider icon to the right of Wi-Fi is turned to blue, if it is not blue, tap on it to make it blue. Tap on Wi-Fi.
5. Find the name of the wireless network you want to connect to. If you are already connected to a network, it will say so beneath the network name.
6. Enter in your wireless network password using the keyboard. (Note: if you do not know your wireless network password, check on the back of your wireless router or cable modem. Sometimes it is listed there.)
7. Tap CONNECT.
8. Your Galaxy S8 will connect to your wireless network.
9. You can press the home button to return to your home screen.

Now that you are connected to Wi-Fi you no longer need to worry about using up valuable cellular data. You also will not need to ever connect to this particular Wi-Fi network again, as your Galaxy S8 will remember this network and connect to it automatically whenever it is in range.

Chapter 5 – Google and Samsung Account

If you read through Chapter 3, then you have already successfully created and signed in to both your Google and Samsung accounts when you first setup your device. In this chapter, I am going to cover the important aspects of both accounts, as well as how to create them and use them.

Google Account

Your Google account is the most important account on your Samsung Galaxy S8 and you need it in order to use your Galaxy S8 fully. Your Google account allows you to download apps, backup certain data, download music and movies, and sync your contacts. I HIGHLY recommend creating a Google Account if you have not done so already.

Please note, that if you have a Google email address such as @gmail.com, then you already have a Google account and do not need to create one. You just need to be sure you are signed in.

How to Create a Google Account or Sign in with your Google Account

1. Open the Apps page from the home screen by tapping down on your screen and swiping up.
2. Find and tap the Settings app.
3. Scroll up or down and find Cloud and Accounts. Tap it.
4. Tap Accounts.
5. If you see Google listed, then you are already signed in with a Google account and do not need to proceed. You can tap on Google to see your account information. If Google does not appear, tap on Add Account.
6. Tap Google.
7. Follow the instructions on your screen to create a Google Account or sign in. See Chapter 3, First-time Setup, step 7 as a reference.

So now that you are signed in to your Google account on your Galaxy, how do you use it? You do not need to worry about that fortunately. Most of the time your Google account will be used automatically, such as when you try to download an app or create your contact list. Your Google account will backup any contacts you create so you never have to worry about losing them. Furthermore, your Google account will allow you to use the Play Store, which is where you can download apps.

Samsung Account

A Samsung account is a little different from a Google account. They both backup important data such as contacts and calendar information. A Samsung account however, is necessary to use certain features on the Galaxy, most notably fully backing up your Galaxy's data and using Bixby. You have probably created and signed in to your Samsung account when you first setup your Galaxy, but just in case, I will show you how to do so.

How to Create a Samsung Account or Login to your Samsung Account

1. From the home screen, tap down on the screen and swipe up to open the Apps page
2. Find Settings and tap it.
3. Tap Cloud and Accounts. If you do not see it, you can scroll up and down to find it.
4. Tap Accounts.

Chapter 5 | Google and Samsung Account

5. If you see Samsung listed, then you have already created and signed in to a Samsung account. In this case, you can tap on Samsung account to view your account information. If Samsung account does not appear, tap on Add account.
6. Tap on Samsung account.
7. Follow the instructions on your screen to sign in or create a Samsung account. Refer to Chapter 3, First-time Setup, step 13.

Once your Samsung account has been created and you are signed in, there is nothing left to be done at this time. You will now be able to use certain features on your Galaxy S8. Furthermore, should you ever get a new phone, some of your data will be saved on your Samsung account and will be easily transferred to your new device as soon as you sign in.

Chapter 6 | Contact List

Chapter 6 – Contact List

Your contact list is an essential tool on your Galaxy S8. With it, you can organize your contacts and store important information, such as a person's phone number, email address, physical address, and more.

Importing Contacts

If your contacts were previously stored on your Google account or Samsung account, then as soon as you signed in with any of those accounts your contacts would have automatically been transferred to your Galaxy S8. If you did not have one of those accounts prior to obtaining a Galaxy S8, then you will either have to import your contacts at your wireless provider's store or create the contacts manually. You can generally import contacts from your old cellular phone by bringing both phones to your wireless provider's store and asking for a contact transfer.

Creating Contacts

Here is how you can create a contact on the Galaxy S8

1. Open the Contacts app from your home screen or from the Apps page. (Swipe up on the home screen to access the Apps page)
2. Tap the large orange + symbol located near the bottom right (Figure 6.1).

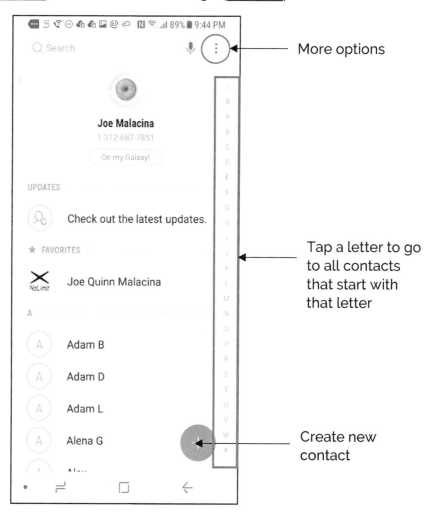

Figure 6.1 – Contacts app

3. Tap into each corresponding box and fill in the contact's information such as Name, Organization, Phone, Email, etc. (Figure 6.2).

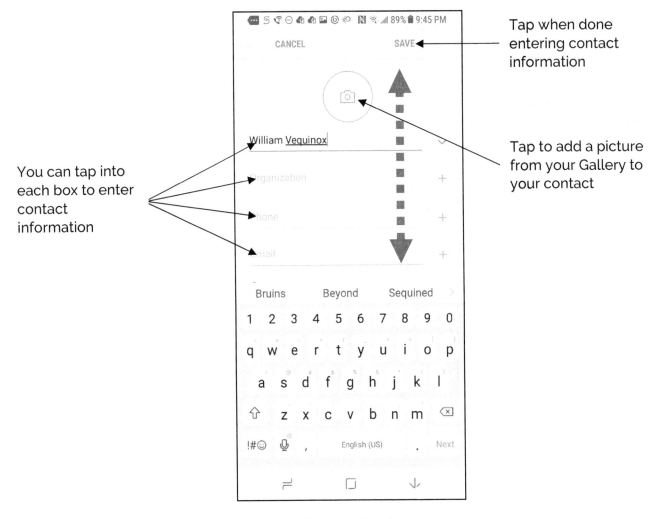

Figure 6.2 – Create New Contact

4. You can scroll down and add further information if you wish.
5. You can tap the camera icon at the top if you want to add a photo from your Gallery for the contact.
6. Tap SAVE at the top when done.
7. Your new contact has just been created.

Chapter 6 | Contact List

Browsing Contacts

You can browse through your contacts in the Contacts app. To view a contact's information, tap on the contact's name, and then tap on Details. You can use the letters on the right hand side of your contact list to quickly browse to all contacts that start with that letter. The FAVORITES group listed in your contact list is generated automatically and is determined by the contacts you interact with most through phone calls and text messages.

Contact Groups

You can create groups for your contacts by tapping on the more options icon (3 vertical dots) and then tapping on Groups (Figure 6.1). From this screen (Figure 6.3), your Galaxy will create some groups automatically, such as company groups and account groups. You can create your own custom group by tapping on CREATE at the upper right. From here, you can name the group, set the group's ringtone, and then add members to the group by tapping on Add member. Once you are done, tap SAVE at the upper right (Figure 6.4).

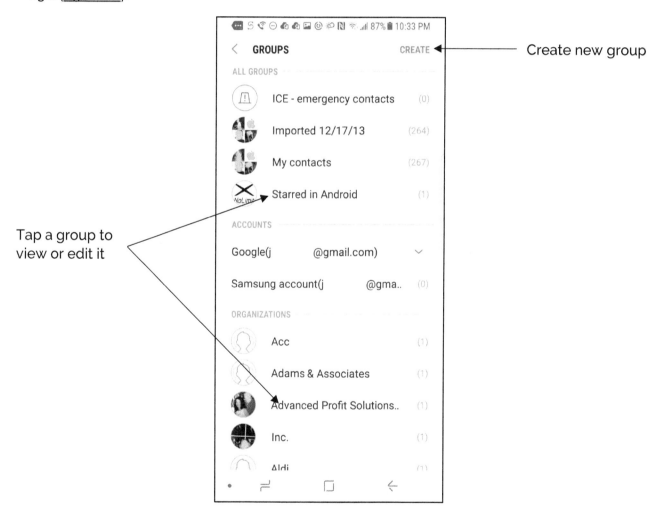

Figure 6.3 – Contacts App -> More Options -> Groups

35

Chapter 6 | Contact List

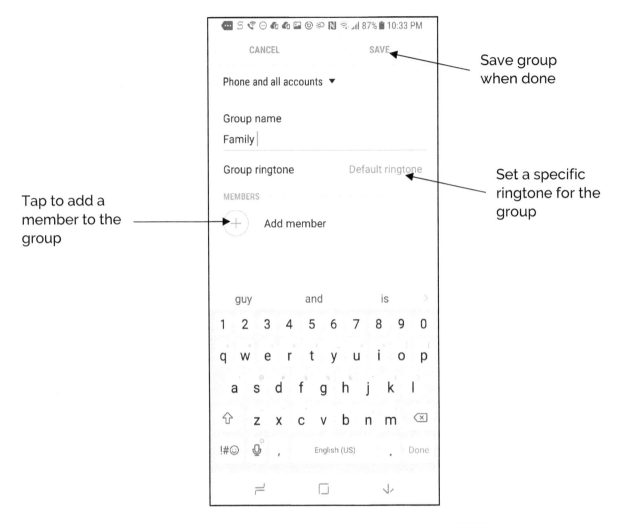

Figure 6.4 – Contacts App -> More Options -> Groups -> CREATE

You can browse through all your groups at the GROUPS page. When you tap on one, you have several options available (Figure 6.5). From here you can see all the members of the group. You can send a group text message or group email using the more options icon at the upper right.

Chapter 6 | Contact List

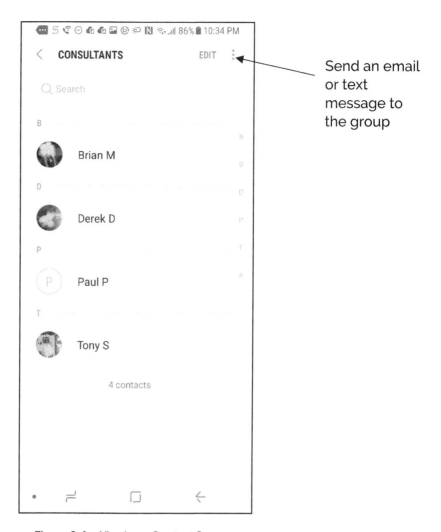

Figure 6.4 – Viewing a Contact Group

Managing Contacts

Managing your contacts through the Contacts app is pretty simple.

How to Delete a Contact

1. Open the Contacts app
2. Find the contact you want to delete and tap and hold down on their name until a checkmark appears
3. Now you can tap as many contacts as you would like to select
4. Tap DELETE at the upper right
5. Tap DELETE again to confirm

How to Edit Contact Information

1. Open the Contacts app
2. Find the contact you want to edit and tap on their name
3. Tap Details

4. Tap <u>EDIT</u> at the top
5. Now you can edit the information of the contact
6. Tap <u>SAVE</u> at the top when done

Chapter 7 | Phone Calls

Chapter 7 – Phone Calls

The Galaxy S8 is a superb cellular phone. Let's explore making and receiving phone calls on the device.

Making a Call

There are numerous ways you can make a call on your Galaxy S8. I will cover the most basic and efficient ways.

How to Make a Call using the Keypad (Figure 7.1)

1. Open the Phone app on you home screen or from the Apps page
2. Tap the green keypad icon at the bottom right

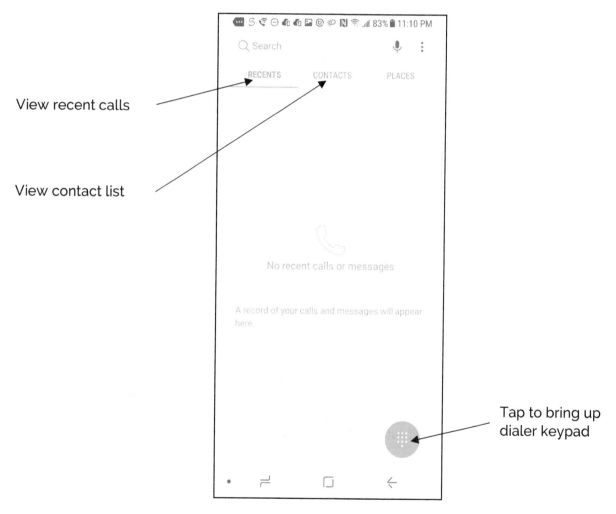

Figure 7.1 – Phone app

Method 1: Dial the Number (Figure 7.2)

Dial the number using the keypad on your screen, and tap the green phone icon to initiate the call.

Chapter 7 | Phone Calls

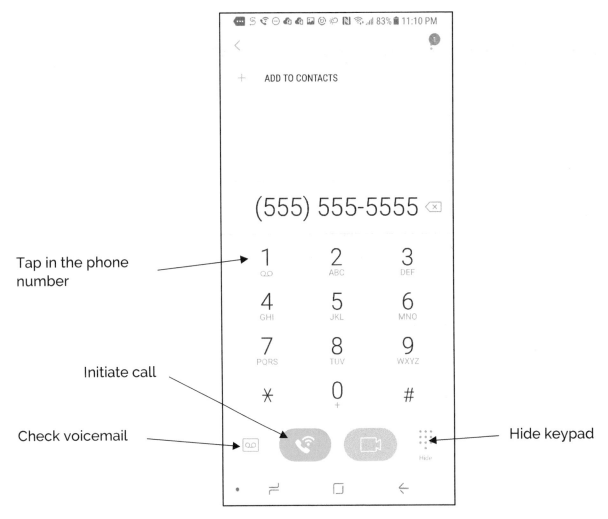

Figure 7.2 – Dialing a Phone Number

Method 2: Dial a few digits of a contact's phone number (Figure 7.3)

This is a quick way to call someone in your contact list of whom you know at least part of their phone number. You can start tapping in the person's phone number and your Galaxy will give suggestions at the top as to who you are trying to call. As soon as you see the person you want to call, tap on their name at the top, and your Galaxy will auto-fill their phone number. Now you can tap on the green phone icon to initiate the call.

After typing in a few digits of a contacts phone number, their name will appear up here. You can tap on their name to auto-fill their phone number

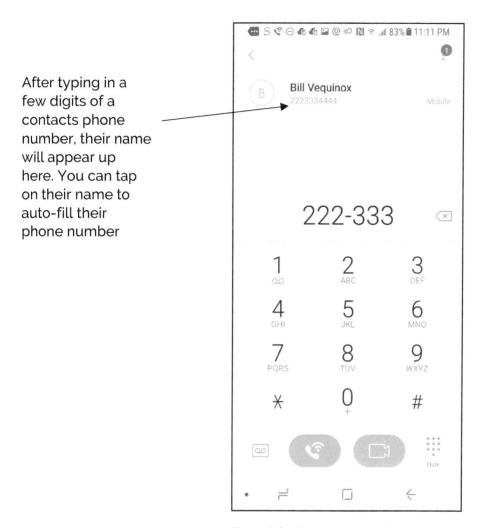

Figure 7.3 – Quick-dial using Phone Number

Method 3: Tap in a contact's name using the keypad (Figure 7.4) *BEST METHOD*

This is similar to method 2 but better. Using the letters on the keypad, you can start typing in a contact's name. You can enter either their first name or last name. As you start typing their name your Galaxy will again give suggestions as to who you are trying to call. You can tap on their name to auto-fill their number and then tap on the green phone icon to initiate a call.

Here is an example of Method 3. Say I wanted to call my friend "Bill Vequinox". Using the keypad, I can tap in the name Bill, using the numbers 2-4-5-5, which spell out Bill. Assuming I do not have a lot of Bills in my contact list, my Galaxy will guess that I am trying to call Bill Vequinox and his name will appear at the top. I can tap on his name which will auto-fill his number and then tap on the green phone icon to initiate the call. A better way would be to just type in V-E-Q, which is 8-3-7. Just this part of his last name will suffice in the Galaxy guessing correctly as to who I want to call. This is by far the quickest and most efficient way to call someone in your contact list.

Chapter 7 | Phone Calls

In this case, you can type in part of a contact's name, and your Galaxy will quickly guess as to who you are trying to call. Again, you can tap on their name when it appears to auto-fill their phone number.

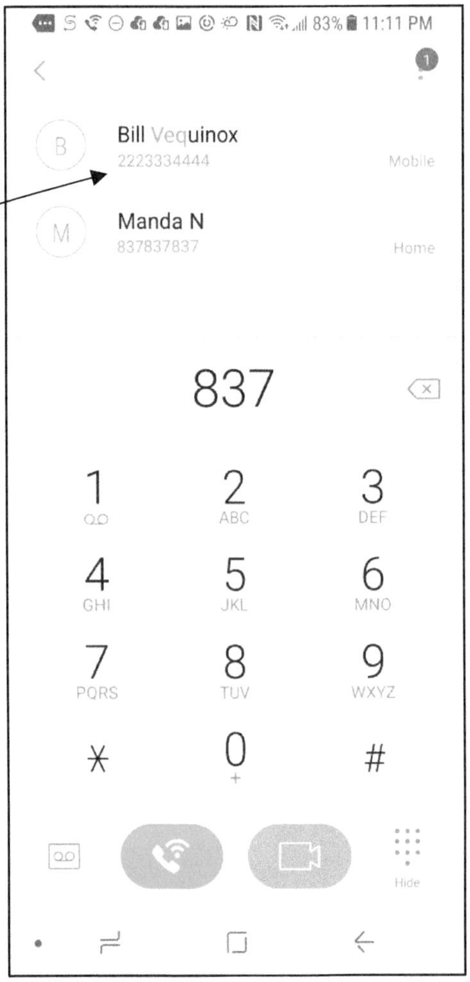

Figure 7.4 – Quick-dial using Contact Name

Receiving a Call
When you receive a phone call, Figure 7.5 will appear. To answer the call, tap down on the green phone icon and swipe to the right. To deny the call, tap down on the red phone icon and swipe to the left. Alternatively, you can tap on the bottom where the words SEND MESSAGE appear and swipe up to bring up some text message options. These options allow you to send a quick text message response to the person who is calling you.

Chapter 7 | Phone Calls

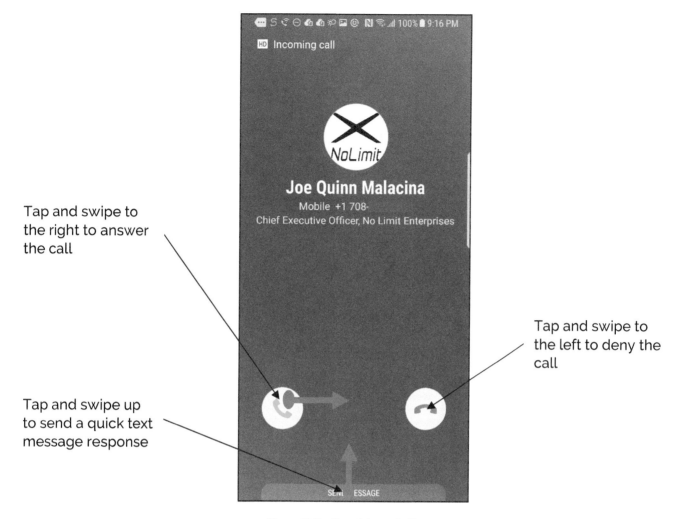

Tap and swipe to the right to answer the call

Tap and swipe up to send a quick text message response

Tap and swipe to the left to deny the call

Figure 7.5 – Receiving a Call

Call Functions

During a phone call, you have several options and controls available to you (Figure 7.6). They do the following if you tap on them:

- **Add Call –** Lets you add another person to the phone call for 3-way calling, if applicable.
- **Video call –** Allows you to convert the phone call to a video call. This is only available if the person you are on the phone with also uses a Samsung Galaxy.
- **Bluetooth –** Allows you to transfer the call to a Bluetooth device such as a headset or vehicle.
- **Speaker –** Turns speakerphone on or off
- **Keypad –** Brings up the keypad
- **Mute –** Mutes your side of the call, i.e. the other person on the line will not be able to hear your end until you turn mute off.
- **Red Phone Icon –** Ends the call

Chapter 7 | Phone Calls

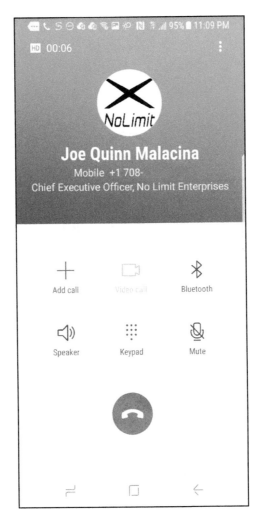

Figure 7.6 – In-Call Functions

Recent Calls

To access your recent calls list, open the Phone app and tap on the RECENTS option at the top. From here, you can view all of your recent calls (Figure 7.7). To call someone on your recent calls list, tap down on their name and swipe your finger to the right. You can also quickly text message someone on your recent calls list by tapping down on their name and swiping your finger to the left. Lastly, to delete calls from your recent calls list, tap down on a call and hold your finger there until it becomes check marked. Now you can tap on additional calls to check mark those as well, or tap on the All bubble at the upper left to select all calls. Then, tap DELETE at the upper right.

44

Chapter 7 | Phone Calls

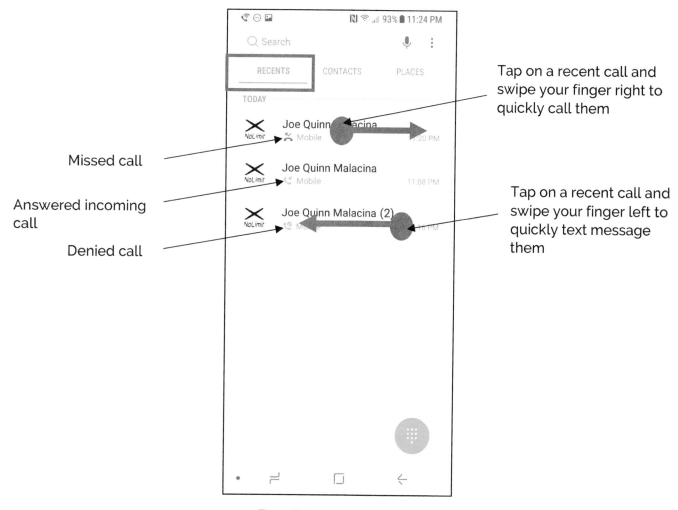

Figure 7.7 – Recent Calls

Voicemail

You can access your voicemail by opening the Phone app and bringing up the keypad. From here, tap on the voicemail icon (Figure 7.2). This will dial your voice mailbox where you can set up your voicemail and listen to messages. Please note that this may not work with all wireless carriers. In fact, some wireless carriers offer apps that allow you to easily check your voicemail right from your home screen. Some carriers even have visual voicemail, which shows your voicemails in text form so you can read them. For more on setting up your voicemail, please contact your cellular provider.

Chapter 8 | Text Messaging

Chapter 8 – Text Messaging

Text messaging on the Samsung Galaxy S8 is a seamless and fun way to communicate with friends and family. Let us explore text messaging on the Galaxy.

Messages App

In order to text message on your Galaxy S8, you will need to use the Messages app. Your cellular provider may place its own text messaging app on your Galaxy S8, but you do not need to use their app. This book will cover the standard Messages app, which is installed on all Galaxy S8 phones. You can locate this app in the Apps page or somewhere on your home screen.

Once you have located the Messages app, open it up. Here in the Messages app is where you will do all of your text messaging. The layout of the Messages app is shown in Figure 8.1.

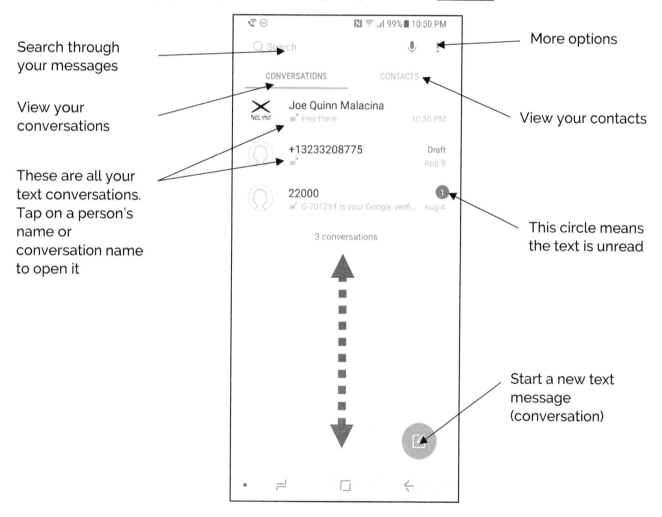

Figure 8.1 – Messages App

Before we explore using the Messages app, let me explain some texting jargon. When you exchange a text message with someone, it is generally called a conversation. In this book, I will refer to any text message communication with another person or group of people as a *conversation*.

47

Chapter 8 | Text Messaging

Sending a Text Message

To send a text message on your Galaxy, tap on the square and pencil icon at the bottom right of your screen within the Messages app (Figure 8.1). This will start a new text conversation. Now your contact list will appear allowing you to browse through your contacts to find someone to text message. When you have found the contact you wish to text message, tap on their name and then tap COMPOSE at the upper right (Figure 8.2). If you are texting someone who is not in your contact list, you can tap into the Search Contacts or enter number box to type in their phone number directly. Lastly, you can browse through your contact groups and recent interactions using the tabs at the top.

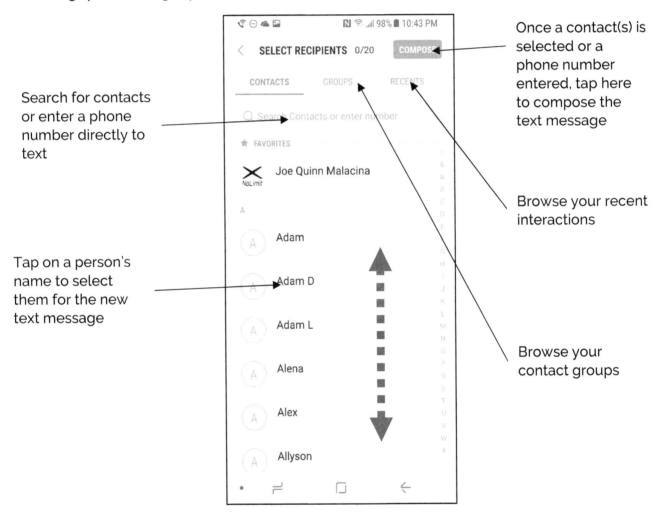

Figure 8.2 – Compose New Text Message

Once you have tapped on COMPOSE, you will be brought to the new text message screen (Figure 8.3). From here you can type your message into the Enter Message box. Once your message is ready to send, tap SEND at the right of the message box. If you need a review on using the on-screen keyboard, see Chapter 4, Figure 4.7.

Chapter 8 | Text Messaging

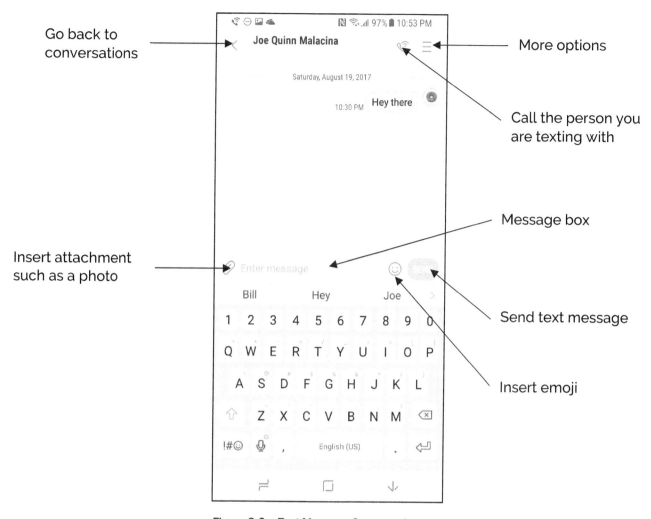

Figure 8.3 – Text Message Conversation

Receiving a Text Message
When you receive a text message, you will receive a notification and the message itself will appear inside the Messages app (See Figure 8.1). Unread messages will have an orange circle icon next to the conversation with a number inside it. To view the message, open the conversation by tapping on the conversation name. Your entire text message conversation will now be shown, including any new messages that you have received in this conversation. You can scroll up and down through the history of the conversation using your finger. At any time, you can return to view all of your conversations by tapping on the back arrow at the upper left.

Group Conversations
With the Galaxy S8, you can hold text conversations with many people at once. The process for doing this is the same as composing a new message, only this time, when you are selecting a person to compose the new text message to (Figure 8.2), select more than one person, and then tap COMPOSE. Now you will be sending a text message to the group of people you selected and this group will be one conversation. You can access the group at any time in the Messages app from the conversations screen.

Chapter 8 | Text Messaging

Attachments

You can send attachments to people through text messages using the paperclip icon (Figure 8.3). From here, you can use the tabs at the bottom to select which type of attachment you want to send. If you tap on the CAMERA tab, your camera will open up allowing you to take a picture right now to send. If you tap on the GALLERY tab, you can browse through your Gallery and select a saved photo to send. Lastly, if you tap on the OTHER tab, you can send a different type of file such as a PDF, Contact, or Calendar appointment. (Figure 8.4)

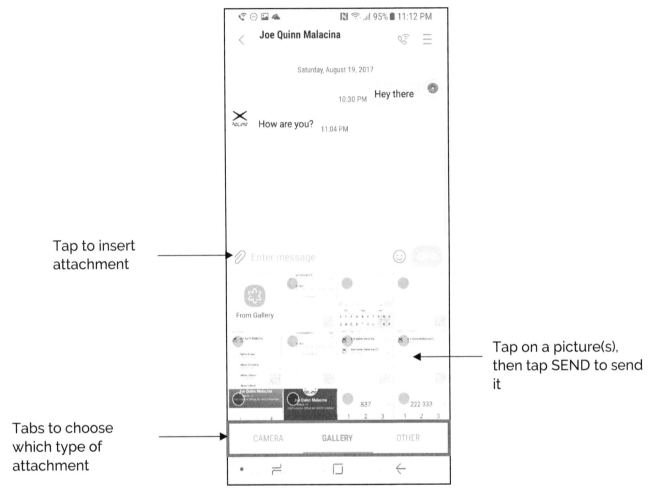

Figure 8.4 – Send Text Attachments

Managing Messages

An important aspect of the Messages app and text messages is how to manage texts and conversations. Let us take a look at a couple of key management features that are prudent to know.

How to Delete Entire Conversations

1. Open the Messages app
2. Tap and hold on a conversation until a checkmark appears to the side of it
3. Now tap to select all conversations that you want to delete

Chapter 8 | Text Messaging

4. Tap <u>DELETE</u> at the upper right
5. Tap <u>DELETE</u> to confirm

How to Delete Individual Text Messages

1. Open the <u>Messages</u> app
2. Tap on a conversation to open it
3. Tap and hold down on an individual message until a popup box appears
4. Tap <u>Delete</u> inside the popup box
5. Now select all the individual text messages you want to delete
6. Tap <u>DELETE</u> at the upper right
7. Tap <u>DELETE</u> to confirm

Please note, at any time you can back out of the selection process by pressing the <u>back button</u>.

Chapter 9 – Email

Your Galaxy has the ability to seamlessly manage your email accounts. It is quite convenient when you can quickly reply to an email in 20 seconds straight from your phone. All of your emails and email accounts are located within the Email app, with you can find on the Apps page.

As a reminder, to access the Apps page, touch down on your home screen and swipe your finger up or down.

Adding an Email Address to your Galaxy

When you first open the Email app, you will be prompted to setup an email account. If you are signed in to your Galaxy with your Google account, the Email app will prompt you to setup this account. To do so, tap on your Google account email address. Your Google email address will set itself up immediately.

You can also add your other email accounts to your Galaxy. To do so, follow these steps:

1. Open the Email app.
2. Tap the 3 horizontal lines at the upper left.
3. Tap the gear icon.
4. Tap Add Account.
5. Enter in the email address you wish to add to your Galaxy and the password to that email address (Figure 9.1).
6. Tap SIGN IN (Figure 9.1).

Chapter 9 | Email

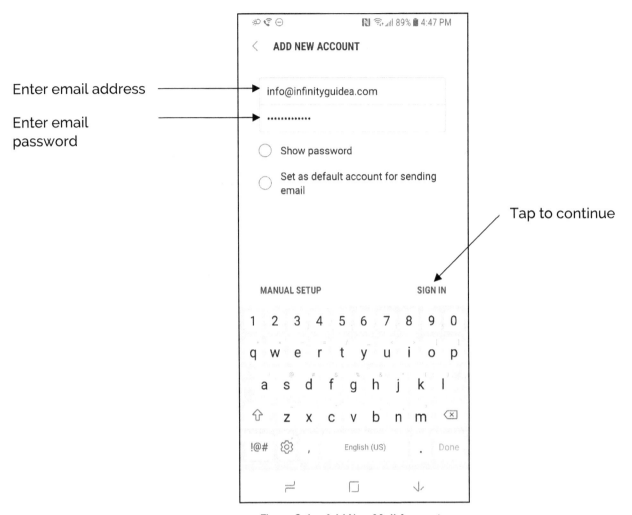

Enter email address

Enter email password

Tap to continue

Figure 9.1 – Add New Mail Account

7. Your Galaxy will attempt to automatically configure your email account to your device. If it cannot do so, a popup box will appear asking for the account type (Figure 9.2). If your Galaxy fails to setup your account automatically, you will need to manually enter your email account's server information. You can generally get this information from your email provider's website or by contacting them. The information you may need include: outgoing server, incoming server, ports, authentications, and security types. For the most popular email accounts such as Yahoo, AOL, Gmail, and Hotmail, your Galaxy should setup your email address automatically.

Chapter 9 | Email

Figure 9.2 – Manual Email Setup

8. Follow the instructions on your screen to finish setting up your email account.

Once your email account has successfully been added to the email app, you are ready to use email on your Galaxy. You can add multiple email accounts to your Galaxy if you wish to do so.

Checking your Email

To check your email, open the Email app. Your email messages will be listed and you can view one by tapping on it (Figure 9.3).

Chapter 9 | Email

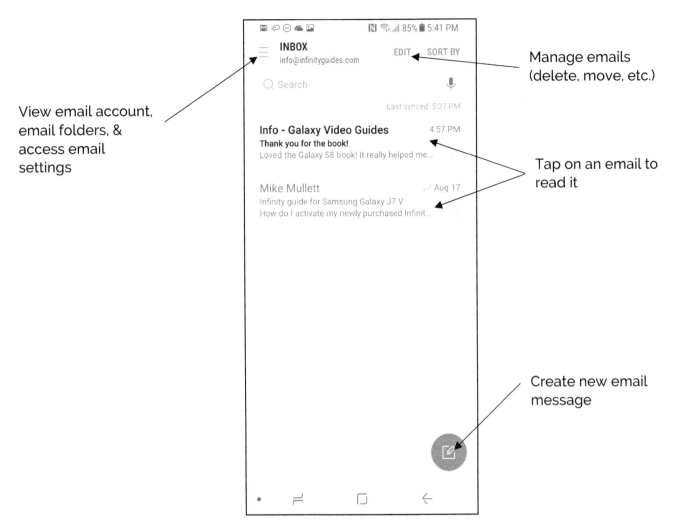

Figure 9.3 – Email App Layout

You can tap the three horizontal lines at the upper left to switch between email account and folders. Simply tap on an account to view your email messages for that specific account (Figure 9.4).

Chapter 9 | Email

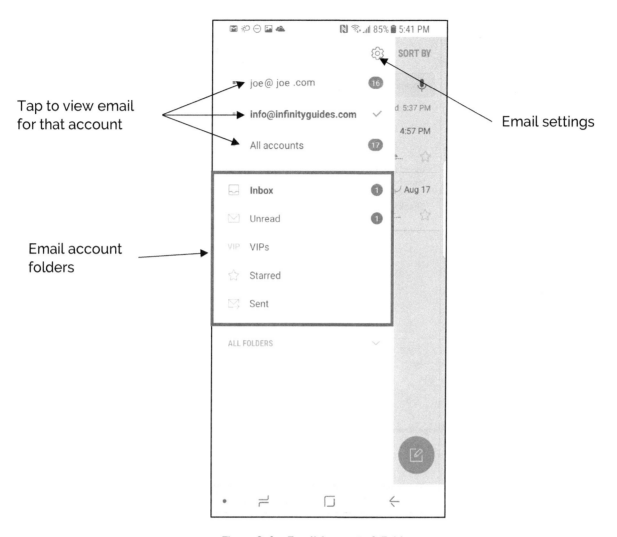

Figure 9.4 – Email Accounts & Folders

Viewing Email

To view an email, simply tap on it and it will appear full screen. From here you have several options available to you including Reply, Reply All, Forward, and Delete. You can use the back arrow or the back button to go back.

57

Chapter 9 | Email

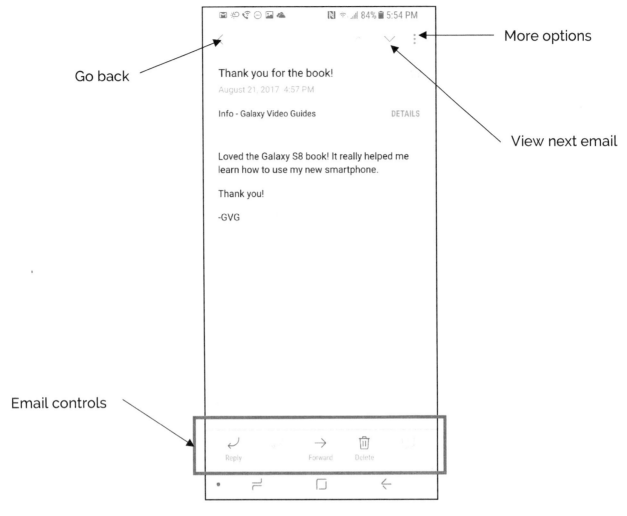

Figure 9.5 – Viewing an Email

Sending Email

At any time from the Email app, you can create a new email by tapping the square and pencil icon at the lower right (Figure 9.3). This will bring up a new email screen (Figure 9.6).

How to Create New Email Message

1. Open the Email app
2. Tap the square and pencil icon at the lower right
3. Tap into the To box and enter in the recipient contact name or email address
4. Tap the From box to change which email address you are sending the email from
5. Tap into the Subject box to enter the email subject
6. Tap into the large text box and enter in the email message
7. To attach a photo or attachment, tap ATTACH
8. Tap SEND at the upper right to send the email

Chapter 9 | Email

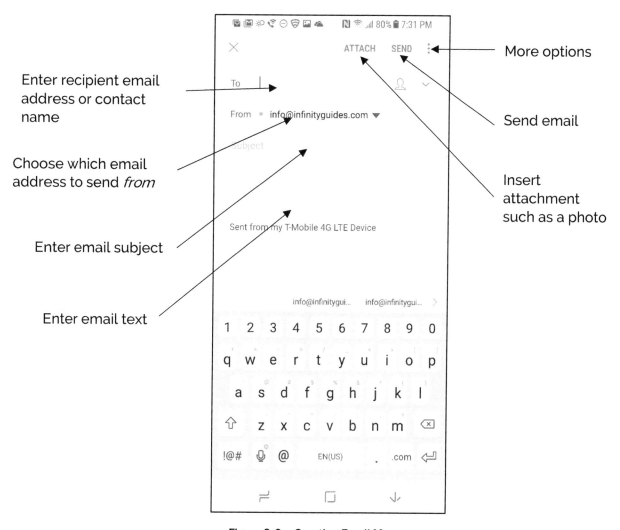

Figure 9.6 – Creating Email Message

Managing Email
How to Delete Email(s)

1. Open the Email app.
2. Tap and hold on an *email preview* until it becomes check marked. (Alternatively, you can tap EDIT at the upper right)
3. Tap each email you want to delete or tap the ALL circle at the upper left to select all emails.
4. Tap DELETE at the upper right.
5. Tap DELETE to confirm.

You can quickly delete an individual email by tapping on it and swiping your finger to the left.

Chapter 10 – Web Browsing

One of the signature features of the Galaxy is the ease in which you can surf the web. To access the Internet, you can use the Internet app or the Chrome app, whichever one you prefer. For simplicity, I am going to show you how to use the Chrome app since it is more widely used. If by some chance your Galaxy does not have the Chrome app (hint: it may be in the Google folder on your Apps page), then this chapter will still work for the Internet app.

Visiting Web Pages

To get started browsing the internet, open the Chrome app from your Apps page or home screen.

At first, Chrome may ask you to sign in with your Google account. If you have followed the advice of this book and are signed in with your Google account, just follow the instructions on your screen until you are brought to the main Chrome page, which is shown in Figure 10.1.

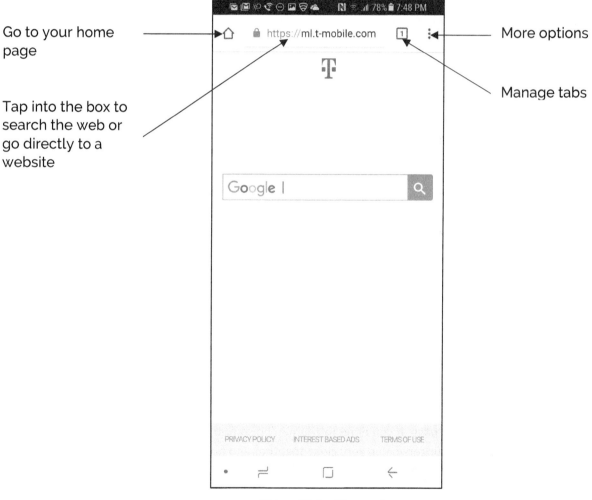

Figure 10.1 – Chrome App

Chrome is Google's web browser optimized for the Galaxy. In Chrome, you can visit web pages as well as search the web. To search the web, tap into the box at the top of your screen and type in your search parameters followed by tapping on GO on your keyboard. If you want to visit a website directly, tap into the same box and type in the web address, followed by tapping GO. You can always return to your homepage by tapping on the home icon at the upper left.

Navigating the web on the Galaxy is a unique experience and takes some practice. Think of your finger as the mouse pointer, and touching down on the screen as a click. To open a link, simply tap down on it with your finger (Figure 10.2).

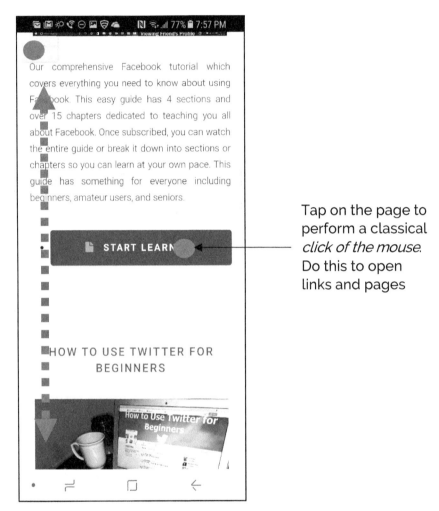

Figure 10.2 - Navigating Web Pages

Scrolling through web pages works the same way as scrolling through apps. Simply drag your finger up or down to scroll through a web page. If you need to zoom in on something, touch down on the screen with two fingers, with your two fingers being very close together and hold them there. Then, separate your fingers while keeping them on the screen. This will zoom in to the center point of your two fingers (Figure 10.3).

Chapter 10 | Web Browsing

To zoom in: touch down on the screen with TWO fingers being close together. Then drag the fingers away from each other in opposite directions while remaining on the screen, releasing at the end or when you are satisfied with the zoom.

Figure 10.3 – Zoom in on a Web Page

To zoom back out, touch two fingers down on the screen again, only this time have the fingers start far apart, and pull the fingers in towards each other (Figure 10.4).

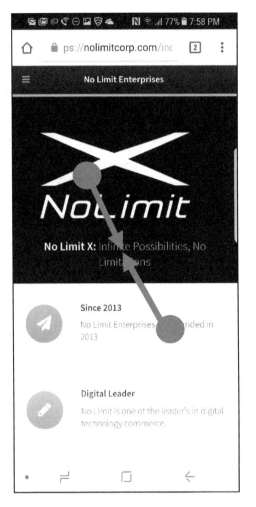

To zoom out: touch down on the screen with TWO fingers spaced far apart. Then drag the two fingers together while remaining on the screen. Release when the fingers come together or when you are satisfied with the zoom.

Figure 10.4 – Zoom out on a Web Page

You can still do all the basic functions of your normal web browser with Chrome. To go back to the previous page, simply press the back button on your Galaxy. You can also manage the web page by tapping on the 3 vertical dots icon which will bring up more options. From here, you can bookmark a page by tapping on the star icon. You can view all of your bookmarks by tapping on Bookmarks. You can set your home page by tapping on Settings and then tapping Home page.

Chapter 10 | Web Browsing

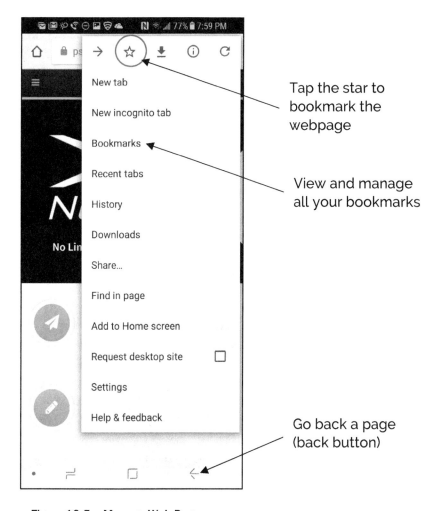

Figure 10.5 – Manage Web Page

Using Chrome Tabs
Chrome offers you the ability to browse multiple websites at the same time using what are called tabs. Tabs are similar to different windows if you were browsing the internet on a computer. You can use tabs to visit several websites at the same time without having to completely close other websites down. You can use tabs in Chrome by tapping on square with the number inside it icon. This will bring up Figure 10.6.

Chapter 10 | Web Browsing

Figure 10.6 – Using Tabs in Chrome

This screen will show you all of the tabs you currently have open. You can quickly go to one of these tabs by tapping on its window or tapping its title. You can create a new browsing tab by tapping on the plus symbol at the upper left. To close a tab, tap the x at the upper right of a tab window or tap on a tab window and swipe it to the left or right off the screen.

Tabs along with the navigational controls discussed in this chapter are essential tools when browsing the internet on your Galaxy. Take some time browsing with Chrome and you will become a pro at mobile internet browsing in no time.

Chapter 11 – Using Your Camera

Your Galaxy is equipped with a powerful camera with flash capability, and is one of the most powerful features of the Galaxy. To access the camera, find and tap on the Camera app on your home screen or Apps page. There are several options available inside the Camera app. On the top of your screen from left to right is: (Figure 11.1)

- **Lens Chooser** – Allows you to switch between the outwards facing lens and the selfie lens.
- **HDR** – Allows you to turn HDR on, off, or auto. HDR stands for High Dynamic Range.
- **Lightning** – Allows you to turn flash to on, off, or auto. A = Auto
- **Gear Icon** – Opens camera settings

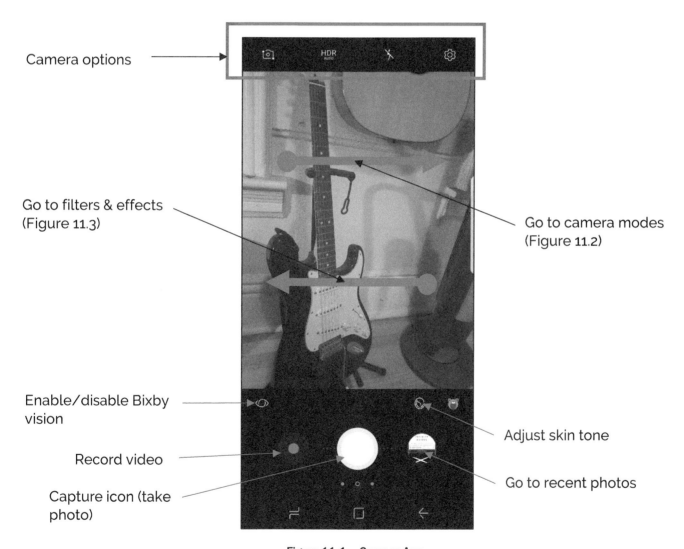

Figure 11.1 – Camera App

As you can see, there is a lot to explore within the Camera app. Let us break it down one at a time.

Chapter 11 | Using Your Camera

Camera Modes

From the Camera app, you can set the camera mode. To do this, tap down on the screen from within the Camera app, and swipe your finger to the right. This will bring you to Figure 11.2. By default, the camera mode will be set to Auto, which is best for most pictures. For the photographically inclined, there are several camera modes you can choose from. To switch the camera mode, simply tap on an option.

You can change the camera mode by simply tapping on an option. Auto mode will be on by default (highlighted)

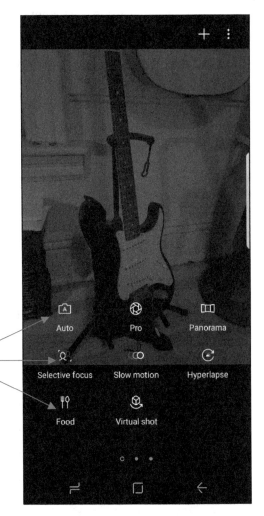

Figure 11.2 – Camera Modes

I will not cover all the different camera modes in this text. For most people, Auto mode will suffice 99-100% of the time when taking a picture. Feel free to explore the other camera modes on your own as some are quite interesting to use. For example, Panorama mode allows you to take a panoramic photo, which is just a very wide photo. Food mode is made for taking pictures of food. Slow motion mode lets you record a video in slow motion. You can even download more camera modes by tapping on the plus symbol at the upper right. You can get back to the main camera screen by pressing the back button.

Chapter 11 | Using Your Camera

Filters & Effects

By tapping down on the main camera screen and swiping to the left, you will be brought to the filters and effect screen of the Camera app (Figure 11.3). All of the different filters are listed near the bottom of your screen, and more are available by swiping left or right over the filters. To enable a filter, tap on it. Other effects you can add to your photo include stickers and facial masks. To remove any effect tap on Remove effect.

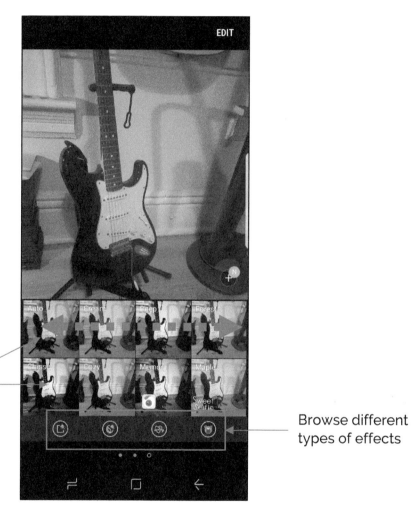

Tap on any filter to enable it

Browse different types of effects

Figure 11.3 – Filters & Effects

Taking a Photo

To capture a photo, tap on the capture icon. Before taking the photo, you can choose to change the camera mode or add a filter or effect. You can also adjust the zoom of the camera by tapping and holding on the capture icon, and then dragging your finger to the left or the right. Lastly, you can adjust the focus of your camera by tapping down on the camera screen where you want the camera to focus. For instance, if I wanted the camera to focus on the base of the guitar in Figure 11.1, I would tap on the base of the guitar on my screen and allow the camera to focus to that before taking the picture. Once the picture is taken, it will appear in the recent photos circle to the right of the camera icon. You can tap on it to quickly go to it.

Recording a Video

To record a video, simply tap on the record video icon instead of the capture icon. Your video will start recording. To stop the recording, tap on the stop icon that replaces the record video icon. Again, your newly recorded video will appear in the recent photos circle when done (Figure 11.1).

Chapter 12 – Photos & Videos

Now that we know how to take photos and record videos, it is time to learn what we can do with these. All of your photos and videos can be found in the Gallery app, which is located somewhere on your home screen or in the Apps page.

Gallery App Overview

When you first open the Gallery app, it may appear as if there is a lot going on (Figure 12.1). Let us take a look at navigating this expansive app. At the top will be your browsing tabs: Pictures, Albums, and Stories. Underneath that will be your content, such as photos or albums. And of course, at the top right will be more options.

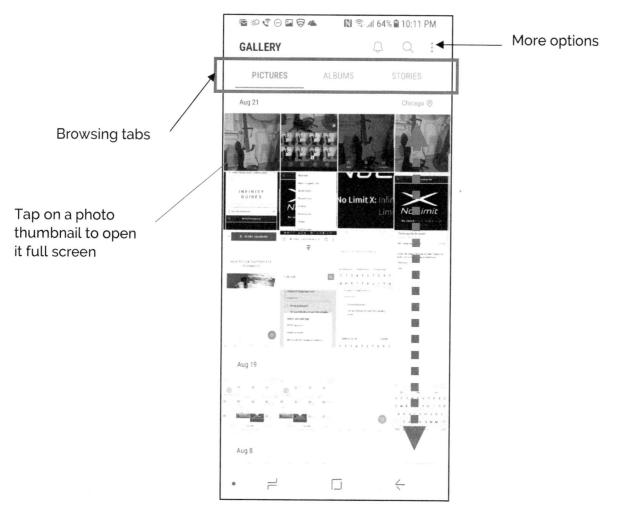

Figure 12.1 – Gallery App

Before we dive in any further, let me give a brief note on photos and videos. Photos and videos take up significant space on your Galaxy. For a lot people, most of the used hard drive space on their phone and SD card is taken up by photos and videos. To save space, you can use what is called *Samsung Cloud*, which I am going to cover right now.

Chapter 12 | Photos & Videos

Samsung Cloud

Samsung Cloud is a cloud based service that allows you to store your photos in the cloud, thus saving space on your phone. The other benefit is if anything were to ever happen to your phone, your photos and videos would be backed up to the cloud for safe keeping. You can activate the Samsung Cloud for free and you will be given a set amount of cloud space you can use. Here's how: (Note: in order to use Samsung Cloud you must have signed up for and be signed in with your Samsung account on your Galaxy)

1. In the Gallery app, tap the 3 vertical dots at the upper right to bring up more options.
2. Tap Settings.
3. Look where it says Samsung Cloud, and make sure the tab mark is enabled on the right. Orange will be shown if it is enabled. Tap on it to enable it.
4. Your Samsung Account email address should be shown underneath where it says Samsung Cloud. If it is disabled and your account email address does not appear, tap on Samsung Cloud to set up the Samsung Cloud service for free. Follow the instructions on your screen.

Browsing Photos

Browsing photos in the Gallery app is simple enough. In the PICTURES tab, you can view all of your pictures. You can tap on any thumbnail to bring the picture to full screen. You can tap on the ALBUMS tab at the top to view all of your photo albums. Some albums will be generated automatically such as "screenshots". Lastly, the STORIES tab is a little different than the rest. Stories are a set of pictures or videos that you can create to be shared with friends who also use Samsung smartphones. Furthermore, inside the Stories tab you can see your friends' stories if they choose to share any. In summary, stories are a unique way of sharing photos daily with friends. I will not cover stories in this text, since there are much easier ways to send and share photos with others.

Photo Albums

One of the best ways to organize and view your photos is through albums.

How to Create a new Photo Album

1. Open the Gallery app.
2. Tap the Albums tab at the top.
3. Tap the 3 vertical dots at the upper right.
4. Tap Create Album.
5. Type in the name of your new album and then tap CREATE.
6. Using the tabs at the top, tap on each picture you want to add to the new album. Each picture will become check marked that is to be added.
7. When done, tap on DONE at the upper right.
8. Your new album has just been created and can be found in the ALBUMS tab.

How to Delete Photos from an Album

1. Open the album you want to edit in the ALBUMS tab.
2. Tap and hold on a picture inside the album until it becomes check marked.
3. Now tap and select all the pictures you want to remove from the album.
4. Tap DELETE at the upper right then DELETE again to remove the photos from your album.

How to Add Photos to an Existing Album

1. Inside the Gallery app, find the photo you want to add to an album.
2. Tap and hold on the photo until it becomes check marked.
3. Tap each photo you want to add to an existing album so they become check marked.
4. Once all the photos you want to add to an album are check marked, tap the 3 vertical dots icon at the upper right.
5. Tap Copy to album.
6. Tap on the album you want to add the photos to.

Sharing Photos

One of the best features of the Galaxy S8 is in the ease in which you can share photos with friends and family. Let us explore how to share a photo with someone.

1. Open the Gallery app and find the photo you want to share.
2. Tap on the photo to bring it up in full screen.
3. At the bottom of the photo should be a horizontal menu, if it does not appear or disappears, tap on the photo to bring it back up (Figure 12.2).

Figure 12.2 - Sharing a Photo

4. Tap on Share.
5. Now you can choose how you want to share it. Tap on the corresponding option (Email, Messages [which is text message], Facebook, etc.) (Figure 12.3).

Tap on any option to share it via that way

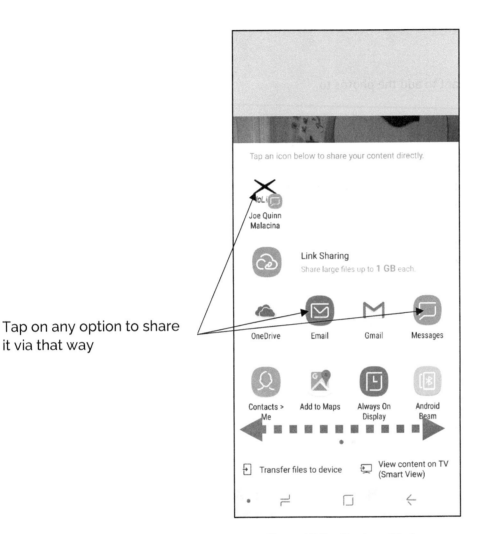

Figure 12.3 – Sharing a Photo

Managing Photos
How to Delete Photo(s)

1. Using the PICTURES tab in the Gallery app, find the photo(s) you want to delete
2. Tap and hold on it until it becomes check marked
3. Tap to select each photo you want to delete
4. Tap DELETE at the upper right
5. Tap DELETE to confirm

Editing a Photo

You can edit a photo in a multitude of ways with the Galaxy. Here's how:

1. Tap on the photo you want to edit in the Gallery app to bring it up full screen.
2. Look for the options at the bottom of the photo. If they do not appear, tap on the photo (Figure 12.2).
3. Tap Edit.
4. You now have several editing options available to you at the bottom of your photo (Figure 12.4).

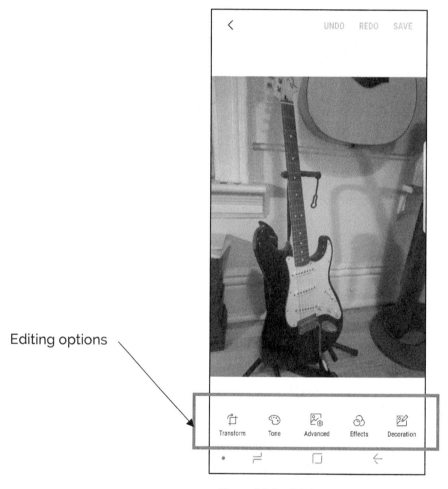

Editing options

Figure 12.4 – Editing a Photo Options

5. The Transform option allows you to edit the physical properties of the photo. This includes cropping, rotating, flipping, and cutting. To crop the photo, drag the border brackets to the desired location (Figure 12.5).

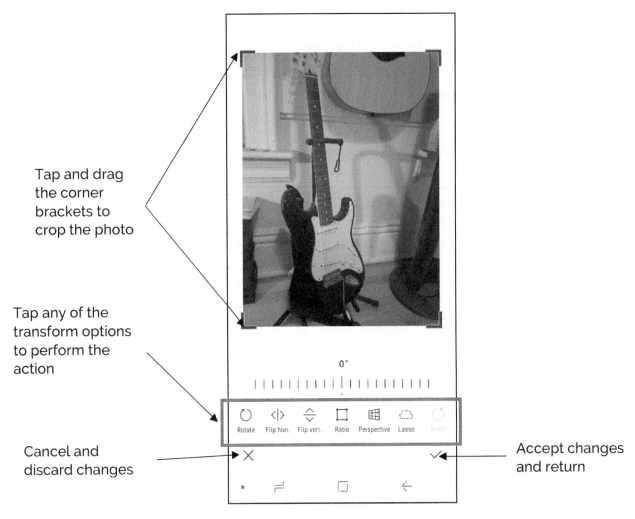

Figure 12.5 – Transform a Photo

6. After performing any changes, tap on the checkmark icon to accept the changes. If you do not want the changes, tap on the x icon to discard the changes and cancel. Both of these icons will bring you back to the main editing options.
7. The Tone option allows you to change the tone and color aspects of the photo. Use the corresponding sliders to change the various properties such as brightness and contrast (Figure 12.6).

Chapter 12 | Photos & Videos

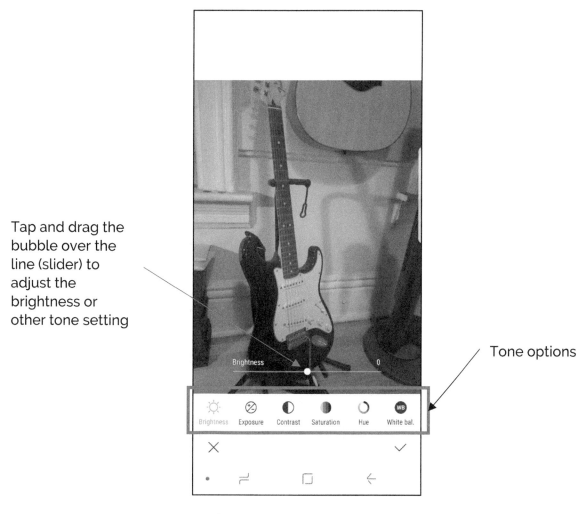

Tap and drag the bubble over the line (slider) to adjust the brightness or other tone setting

Tone options

Figure 12.6 – Tone a Photo

8. The Advanced option allows you to alter advanced properties of the photo.
9. The Effects option allows you to add a filter to the photo. Simply scroll and tap on the filter you want to apply to the photo. Remember, tap on the checkmark to accept a change or the x icon to discard a change (Figure 12.7).

Figure 12.7 – Add a Filter to a Photo

10. The Decoration option allows you to decorate the photo. You can superimpose another image onto the photo, place a sticker on it, label the photo, cover the photo, and draw on it (Figure 12.8).

Chapter 12 | Photos & Videos

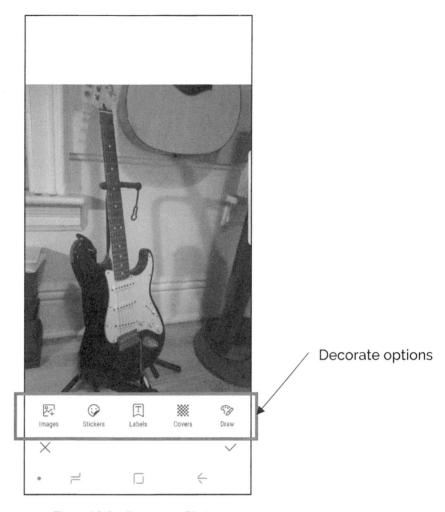

Figure 12.8 – Decorate a Photo

11. If you <u>Draw</u> on the photo, you can choose the color and size of the pen by tapping on <u>Pen</u>. Then to do the drawing, use your finger and draw on the screen (<u>Figure 12.9</u>).

Chapter 12 | Photos & Videos

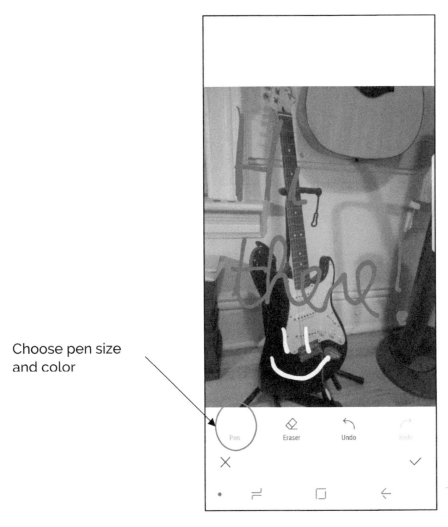

Choose pen size and color

Figure 12.9 – Draw on a Photo

12. When you are done editing a photo, tap on the checkmark to get back to the main editing screen, and then tap SAVE at the upper right (Figure 12.10). Your edited photo will be saved in your Gallery.

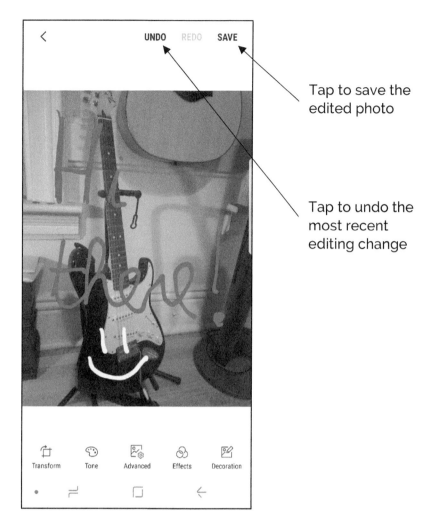

Figure 12.10 – Save an Edited Photo

Chapter 13 – Galaxy Security

Security is a very important feature of the Galaxy. Many Galaxy phones have multiple security features you can enable. In this Chapter, I will cover some of the most important security features for your device.

Settings

All of your security features can be viewed and customized inside the Settings app. There are two different ways you can access Settings on your Galaxy. The first, and most basic way is to find it inside your Apps page. It will appear as an app just like any other. The second way is to access Settings from the Notification Bar. The Notification Bar will be covered fully in Chapter 18, but for now I will show you how to access Settings from it.

To first access the Notification Bar, tap down at the top of your home screen and swipe down. This will bring up what is known as the Notification Bar. Now find the gear icon that is located at the upper right, and tap it to access Settings (Figure 13.1).

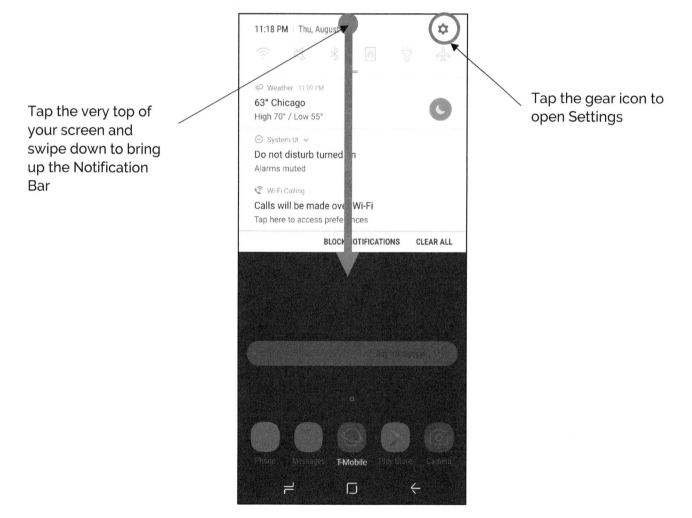

Figure 13.1 – Access Settings through the Notification Bar

Chapter 13 | Galaxy Security

Here inside the Settings app is where you can customize all the under-the-hood aspects of your Galaxy. We will explore more in Settings later on, but for now, locate Lock screen and security and tap on it (Figure 13.2).

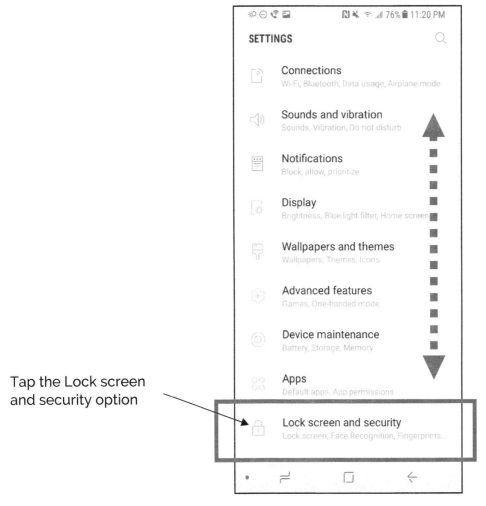

Tap the Lock screen and security option

Figure 13.2 – Settings

Now that we are inside Lock screen and Security, we can explore all the different general security options for your Galaxy. The most essential security you should set up is screen lock. Screen lock is security that prevents anyone from using your Galaxy without a proper password or authorization. Your lock screen itself is just the first screen you see when you power on the screen of your device. If you tap on Lock screen type, you can see the different options:

- **Swipe** – To get past your lock screen and into your home screens you will have to swipe your finger on the screen. This option provides no security whatsoever.
- **Pattern** – Your Galaxy will require you to draw a simple pattern in order to unlock it.
- **PIN** – Your Galaxy will require you to enter a PIN number in order to unlock it.

- **Password** – Your Galaxy will require you to enter a password in order to unlock it.

The above security options are listed from least secure to most secure. For most people, Pattern security will work just fine. If you want the most security possible, choose Password. I will demonstrate how to set up Pattern security in just a moment.

In addition, the Galaxy S8 offers biometric security. These options are less reliable but are still a viable option:

- **Face** – Your Galaxy will scan and read your face thereby only unlocking when you are looking at the front camera of the Galaxy.
- **Fingerprints** – With this option, your Galaxy will require your fingerprint in order to unlock your device. The fingerprint scanner is on the back of the Galaxy next to the camera.
- **Iris** – Lastly, the Iris security option scans the Iris of your eye and uses that to unlock your device.

In order to setup biometric security, simply tap on the option and follow the instructions on your device to set it up. You will also be required to setup standard non-biometric security, which I will illustrate now.

How to Set a Pattern as Screen Lock Security or Change Lock Screen Security

1. Open the Settings app.
2. Tap Lock screen and security.
3. Tap Screen lock type.
4. If you have setup screen lock security before, you will be asked to enter it in now to proceed. Do so.
5. Tap on one of the top options to enable security of that type. In this case, tap on Pattern.
6. Using the dots that appear on the screen, draw a pattern with your finger, with your finger remaining on the screen until your pattern is drawn. In this instance, we are drawing a letter M (Figure 13.3).

Chapter 13 | Galaxy Security

To create your security pattern, tap down on a dot on the screen, and drag your finger to another dot, followed by another, and another, until you draw a pattern that you can remember. When you are satisfied with your drawn pattern, release your finger. In this example, I drew the letter "M" using the dots; something that is easy to remember for me

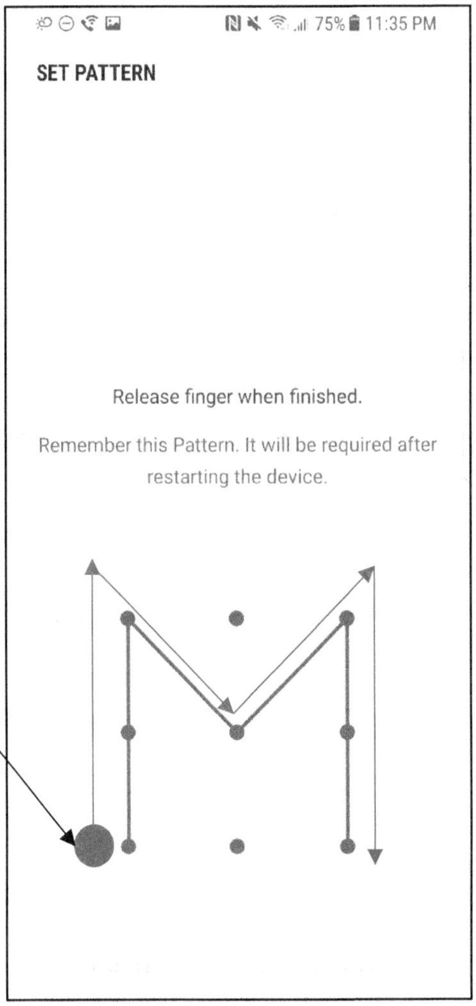

The order in which you draw the pattern matters. For instance, in this example I started with the dot at the lower left to draw an "M". If after setting this pattern I tried to draw an "M" starting at the lower right, the pattern would not work and I would not be able to get through my security. Remember where you start your pattern and where you end it!

Figure 13.3 – Creating a Lock Screen Security Pattern

7. Tap CONTINUE at the bottom.
8. Draw your pattern again to confirm it.
9. Tap CONFIRM at the bottom.
10. Your pattern will now be set.

To test your new pattern, turn off your Galaxy's screen by pressing the power button and quickly releasing. Then, turn the screen back on again and swipe your finger over the screen (Figure 13.4).

Chapter 13 | Galaxy Security

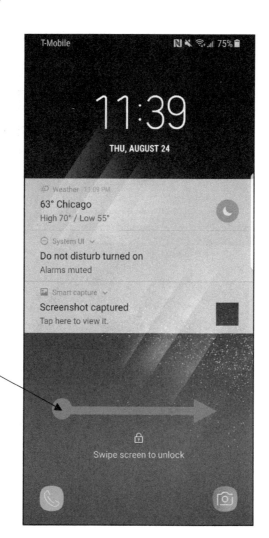

To proceed past your lock screen, tap down on the screen and swipe your finger across

Figure 13.4 – The Lock Screen

In order to proceed your Galaxy will ask for your security pattern (Figure 13.5). Draw your pattern with your finger and your device will unlock and bring you to your home screen.

Chapter 13 | Galaxy Security

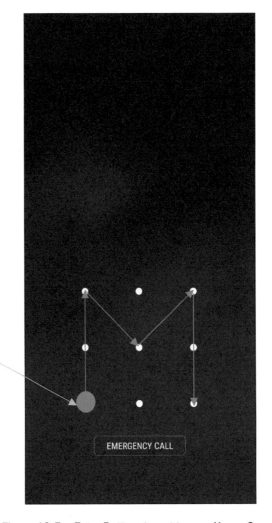

Draw your pattern exactly the way you set it in order to proceed to your home screen and "unlock" your device

If you have enabled lock screen security on your Galaxy, you nor anyone else will be able to proceed to your home screen until your pattern is drawn or whatever security type you chose

Figure 13.5 – Enter Pattern to get to your Home Screen

You have now successfully set up lock screen security on your Galaxy. I highly recommend writing down your lock screen pattern, PIN, or password somewhere safe as you will NOT want to forget it. Getting into your device if you forget your lock screen security is a real headache. If you do not want any lock screen security, simply follow the steps in *How to Set a Pattern as Screen Lock Security or Change Lock Screen Security*, and in step 5, choose Swipe.

Other Security Settings
There are other security settings you can explore in Settings -> Lock screen and security. These are more advanced and for most people not required. Some notable options are SD card encryption, Secure Folder, and Find My Mobile. Browse these at your leisure.

Chapter 14 | Personal Settings

Chapter 14 – Personal Settings

There are a number of personal settings you can customize on your Galaxy, and in this chapter we will explore some of them.

Setting your Wallpapers

Here is how to set the wallpaper, or background image on your Galaxy:

1. While on your home screen, tap down on a blank area of the screen and hold your finger there until your screen splits apart (Figure 14.1).

Tap and hold on a blank area of your home screen until your screen "breaks apart" and Figure 14.2 appears

Figure 14.1 – Enter the Home Screen Editor

2. Tap on Wallpapers and themes (Figure 14.2).

89

Chapter 14 | Personal Settings

Tap to customize your wallpapers and themes

Figure 14.2 – The Home Screen Editor

3. If a message appears, tap ALLOW, otherwise continue to step 4.
4. Figure 14.3 will appear. To browse the default wallpapers on your Galaxy, tap on the words VIEW ALL.

Chapter 14 | Personal Settings

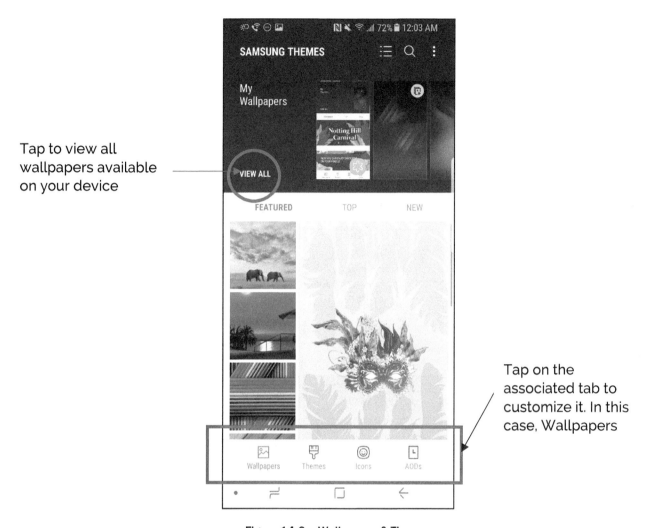

Tap to view all wallpapers available on your device

Tap on the associated tab to customize it. In this case, Wallpapers

Figure 14.3 – Wallpapers & Themes

5. On this screen, you can browse and select a wallpaper by tapping on it. To choose a picture from your Gallery, tap on the square that says From Gallery (Figure 14.4). Once you have found the image you want, tap on it.

Chapter 14 | Personal Settings

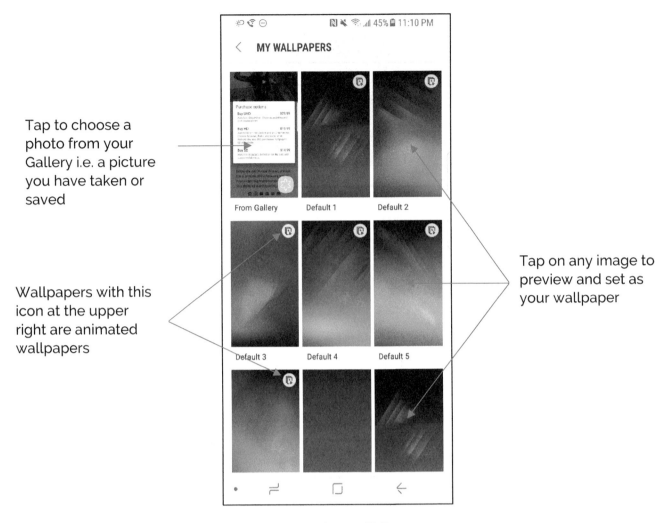

Figure 14.4 – Select a Wallpaper

6. Tap on whether you want this wallpaper to be applied to your Home screen, Lock screen, or both.
7. Tap on SET AS WALLPAPER.
8. Your new wallpaper will now be set.

You can also download wallpapers from within the Wallpapers and themes screen.

Themes & Other Visual Customizations

Besides wallpapers, you can customize some other visuals on your Galaxy. The largest customization is the theme. Themes are everything, which include icons, color schemes, and wallpapers. I would caution against downloading a new theme unless you are very well accustomed to using your Galaxy. This is because a new theme can make it seem harder to find certain tasks and functions since everything will look a little different.

Chapter 14 | Personal Settings

Another aspect you can customize is the Always on Display (AOD) and the appearance of icons. The AOD is the main screen you see when your Galaxy's screen is off. The AOD, icons, and themes can be customized the same way as wallpapers by using the tabs in Figure 14.3.

Please note that themes and wallpapers that are not pre-installed on your Galaxy may cost money to download.

Sounds & Ringtones

Many people like to customize the ringtone on their device. Let's explore how.

How to Set or Change Your Ringtones

1. Open the Apps Page on your device by tapping on Apps or tapping on your home screen and swiping up.
2. Tap Settings.
3. Tap Sounds and vibration.
4. Tap Ringtone (Figure 14.5).

Figure 14.5 – Settings -> Sounds and Vibration

5. Browse through the list of ringtones and tap on one to listen to it. When you are satisfied with the selected ringtone, use the back button to go back and your ringtone will be set (Figure 14.6).

Chapter 14 | Personal Settings

Your last selection will become your ringtone. Use this arrow to go back

Tap on a ringtone to preview it through your Galaxy's speakers

Figure 14.6 – Settings -> Sounds and Vibration -> Ringtone

There are many sound settings you can customize from the Sounds and vibration screen inside Settings (Figure 14.4). Browse these at your leisure and tap on an option to customize it. You can also customize the vibration settings of your Galaxy including the vibration pattern and intensity.

Chapter 15 – Apps

Everything the Galaxy can do is a function of apps. When you make a phone call, you are using the Phone app. When you send a text message you are using the Messages app. Even when you are changing your settings, you are using the Settings app. The apps that come with the Galaxy are incredibly useful, but these are just a needle in haystack compared to the vast amount of apps available to you. To get new apps, you will need to use an app called the Play Store, which is located somewhere on your home screen, or in the Apps page.

The Play Store

Let's open the Play Store app by bringing up the Apps page and tapping on its icon. In order to use the Play Store, you must have and be signed in with your Google account. If you are not, the Play Store will prompt you to sign in or sign up. If you already have a Google account and are not signed in, go ahead and sign in. For more information on creating and using a Google account, please refer to Chapter 5.

There are two main types of content you can download from the Play Store. The first is apps, which are applications you can use on your Galaxy. Apps are very wide ranging and can include games, productivity tools such as word processors, video players, and much more. Many apps are free and some cost money to download. Of the apps that are free, some offer what are called in-app purchases. These are things that you can purchase directly from the app, such as premium rights or special features. The second type of content is media. Media can include anything such as music, movies, television shows, eBooks, magazines, and more. Comparatively to apps, some content is free and others cost money to download.

Browsing the Play Store

Here in the Play Store you can browse through the massive library of apps available for download (Figure 15.1). You can also use the Play Store to download music, movies, and other media. When using the Play Store, make sure you remember your Google account password. It may come in handy.

Here at the main page of the Play Store, called the Apps & Games page, you can see all the featured apps that Google recommends. Let us explore the layout of this page, and refer to Figure 15.1.

Chapter 15 | Apps

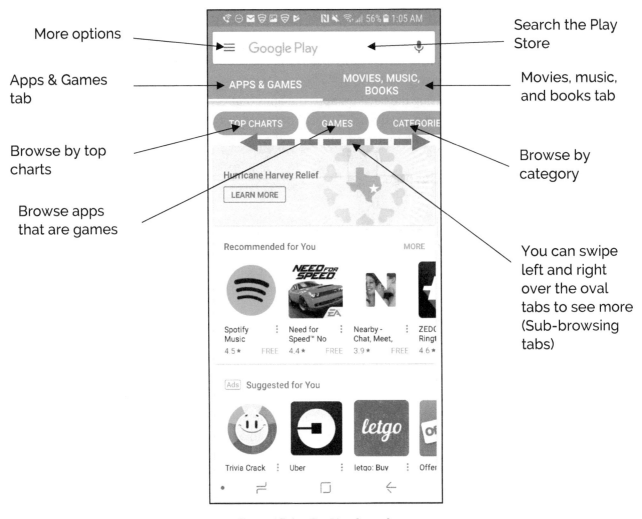

Figure 15.1 – The Play Store App

Main Browsing Tabs

- **APPS & GAMES** – Browse through all the apps and games available for download in the Play Store
- **MOVIES, MUSIC, BOOKS** – Browse through media content available in the Play Store such as movies, television shows, and books

Sub-browsing Tabs for Apps & Games

- **TOP CHARTS** – Browse apps by popular metrics such as top free apps, top free games, top grossing apps, trending now apps, top paid apps (cost money to download), and top paid games
- **GAMES** – Browse through downloadable games with subcategories of its own
- **CATEGORIES** – Browse categories of apps, such as educational, video, cooking, and more.
- **EDITOR'S CHOICE** – See which apps are currently recommended by the editors of the Play Store
- **FAMILY** – Family friendly apps
- **EARLY ACCESS** – See brand new and upcoming apps

Chapter 15 | Apps

Browsing by Top Charts

One of the best ways to find a good app to download is to browse by top charts. To do this, tap on APPS & GAMES inside the Play Store and then tap on TOP CHARTS (Figure 15.1).

From here, you can use the tabs at the top to browse through the top charts (Figure 15.2).

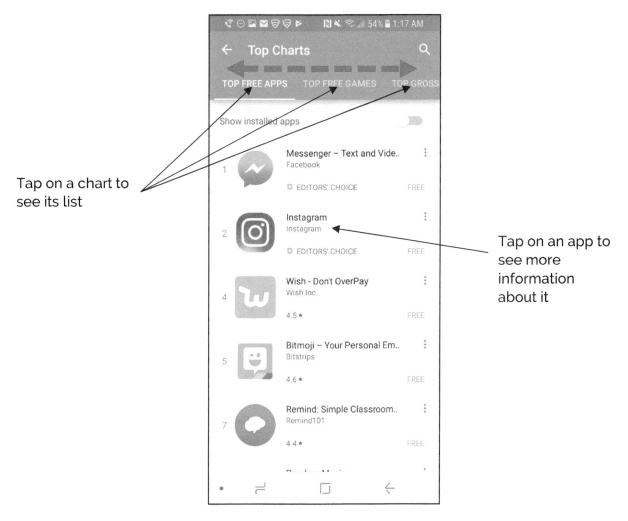

Figure 15.2 – Play Store -> Apps -> Top Charts

If you tap on TOP FREE APPS, this will show you the most downloaded and used apps that are free to download. You can tap on an app to learn more about it (Figure 15.3).

97

Chapter 15 | Apps

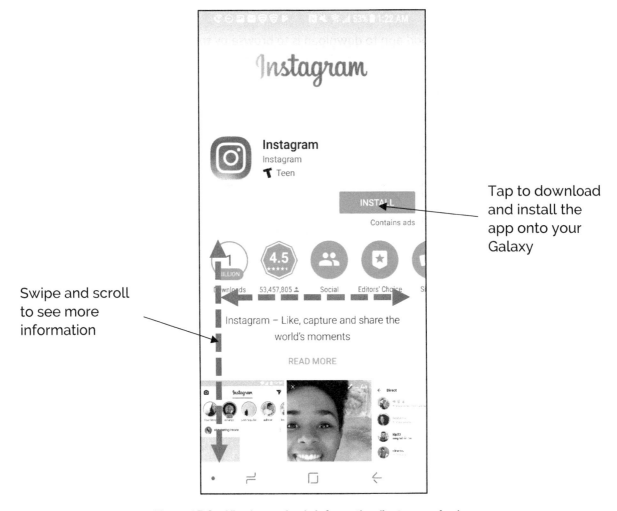

Figure 15.3 – Viewing an App's Information (Instagram App)

From here you can scroll up and down to read a brief description of the app, see screenshots, read user reviews, and more. To download an app, tap on the INSTALL icon.

Also within top charts, you can browse through the top games, top grossing apps, and top apps that cost money to download.

Browsing by Category

Another great way to find good apps to download is to browse by category. To browse by category, tap the CATEGORIES tab under APPS & GAMES. From here you can swipe up and down to view various categories, and you can tap on a category to be brought to that category's top charts (Figure 15.4). You can tap on an app to view more information about it.

Chapter 15 | Apps

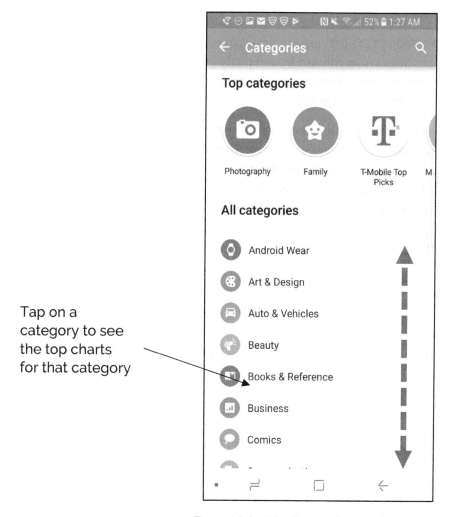

Tap on a category to see the top charts for that category

Figure 15.4 – Play Store -> Apps -> Categories

Searching for an App

You can search for an app from within the Play Store by using the search tab at the top (Figure 15.1). Here you can type in the name of an app you want to explore, or you can type in a keyword for an app such as photography or books. When you execute the search, the Play Store will show a list of apps, and you can tap on app to view more information.

Downloading Apps

To download an app from the Play Store, browse or search for the app you want to download and tap on it to bring up more information. Then tap on the INSTALL icon. Another prompt may appear which you will have to tap Agree. The app will start downloading and when it is finished it will appear on your Apps page.

To use a newly downloaded app, simply tap on it from within the Apps page and enjoy. Using downloaded apps is much like using any of the apps we have covered so far in this book. You can usually scroll through screens on the app, tap icons, and more. Each app is unique to itself, so you will have to explore it to become accustomed to using it.

Important Information about Downloaded Apps

There are certain aspects of downloaded apps that are important to consider. The first is in-app purchases. Many apps allow you to buy certain content from directly inside the app. These purchases could be access to additional content or special features such as removal of advertisements. To make these purchases you will have to use your Google account.

Another aspect to consider is advertisements within apps. For most apps these are inconspicuous and do not interfere with using the app, but for some apps the advertisements can be quite bothersome. If an app is flooded with advertisements and popups that make the app difficult to use I would consider deleting the app. Apps like these may even offer a paid version that will remove the advertisements. I personally recommend not buying the ad-free version since apps like these may just be trying to spam you with ads rather than provide a good service. On the other hand, if an app shows advertisements that does not interfere too much with its use and offers an ad-free version of the app for a small price, I would consider purchasing it if you use the app often.

The final aspect to consider when downloading apps is security. Google does an excellent job at rooting out bad apps and keeping them safe in the Play Store but some slip through the cracks and can cause problems. These apps may phish for your information or cause your Galaxy to slow down. A good practice to utilize would be to read the reviews of any app that you have not heard of before to make sure it is safe to use. If you see a lot of reviews complaining that the app crashes often or is a scam, do not download it.

I have provided a list and brief description of excellent apps you can download on your Galaxy in the Appendix of this book. The apps I recommend are popular, reputable, and for the most part, easy to use. You can also learn more about using apps at www.infinityguides.com. The website provides tutorials for popular apps on the Galaxy and other devices.

Downloading Media Content

You can browse and download media content much the same way you can for apps. In the Play Store, tap on MOVIES, MUSIC, BOOKS at the top of your screen. This screen will show you featured media content (Figure 15.5), and you can tap on any content to view more information (Figure 15.6). Some content is free to download while others you must purchase.

Chapter 15 | Apps

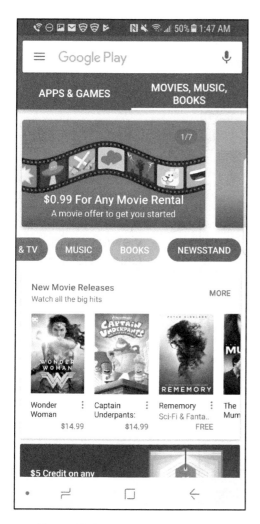

Figure 15.5 – Play Store -> Movies, Music, Books

Figure 15.6 – Viewing Media's Information (Wonder Woman Movie)

Any media content you download or purchase from the Play Store on your Galaxy will become available through your Google account. This basically means that if you purchase a movie such as *Wonder Woman* on your Galaxy, you can choose to watch it on nearly any device that you can sign in to your Google account from including your Galaxy phone or tablet, computer, Smart TV, Chromecast, or other media device. The same goes for eBooks, television shows, and purchased music.

When you go to purchase media content, the Play Store will usually tell you how the media can be played (Figure 15.7). Your Galaxy S8 can play movies and television in full HD quality and does so from the Google Play Movies & TV app. Downloaded music will be played on the Play Music app. If you download content that your Galaxy cannot play immediately, such as an eBook, your Galaxy will tell you that you need to download an app to play it, and may direct you to download a specific app such as the Play Books app to read books.

101

Chapter 15 | Apps

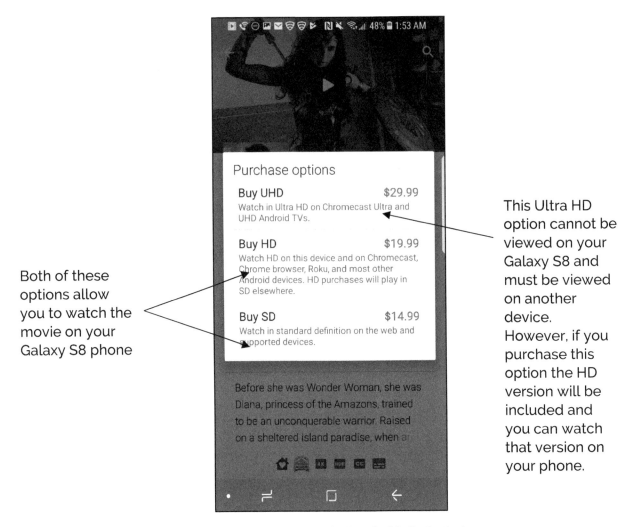

Figure 15.7 – Purchase Options for Media Content

Once you download media content or an app, it is usually yours to keep forever. If you delete it, you can always get it back for free from the Play Store. This does not necessarily apply if the media content was rented or if the content becomes no longer available in the Play Store.

Additional Resources for Play Store Apps & Media Content
The Play Store and the vast amount of apps and content available within it make it impossible to cover thoroughly in one book. Luckily there are additional resources available to help you learn how to use certain apps and to keep up on the latest info. For learning individual apps I would recommend visiting www.infinityguides.com. That website has some good tutorials on popular apps such as Facebook, Snapchat, Twitter, and Instagram. For general information on smartphone technology you can visit www.dummiesblog.com, which often has important news and information on Samsung phones and the Play Store.

Chapter 16 – The Home Screen

By now, you should be pretty familiar with your home screen. As a reminder, your home screen is the main screen of your Galaxy that shows your apps and widgets. You have multiple home screens, and you can switch between them by tapping down with your finger on the screen and swiping left or right. To get back to the main home screen at any time, simply press the home button. (See Figure 4.1 in Chapter 4 as a reminder)

Apps

Your home screen is the place where you can access your apps. Apps are the oval-like icons on your home screen. To view all the installed apps on your Galaxy, you need to access the Apps page. You can do this by tapping down on the home screen and swiping your finger up. (See Figures 4.1 & 4.2 as a reminder. To open an app, simply tap down on it with your finger.

Widgets

Widgets are interactive snippets of an app that you can place on your home screens. They can do a number of things such as display information or allow you to browse. An example of a widget on your main home screen is the area where it shows the current temperature (Figure 16.1). This is the weather widget and if you tap on it, your Galaxy will open The Weather Channel app. Another widget on your home screen is the Google search bar. Tapping on this widget will open the Google app and allow you to search the web.

Chapter 16 | The Home Screen

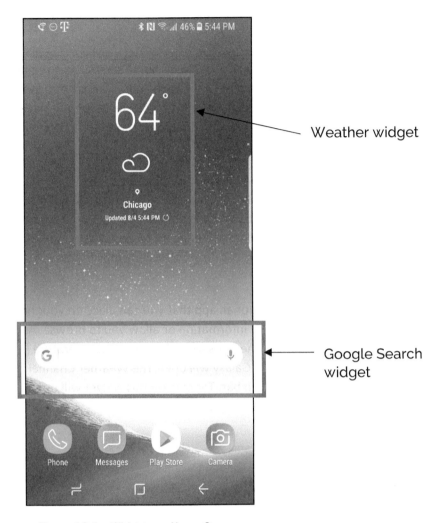

Figure 16.1 – Widgets on Home Screen

Organizing Home Screens, Apps, & Widgets

There is a lot you can do with your home screens on your Galaxy. You can choose which apps appear and where they appear, along with place and edit the size of widgets. You can also create multiple home screens. To start editing your home screens, tap down on the home screen and hold until your home screen breaks apart and Figure 16.2 is shown.

Chapter 16 | The Home Screen

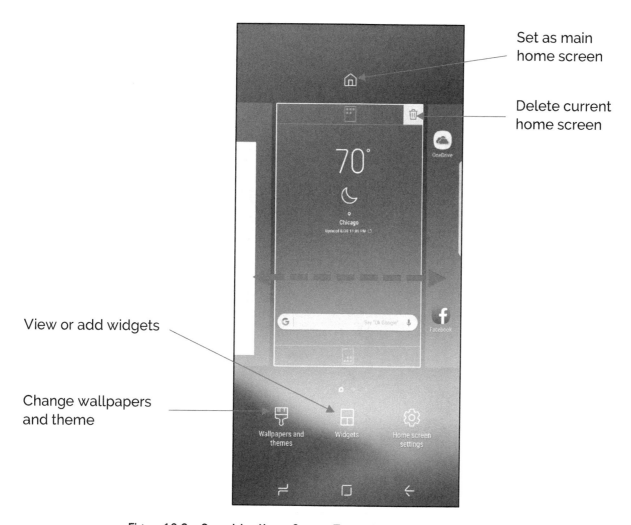

Figure 16.2 – Organizing Home Screen (Tap and hold on blank space of home screen)

From this screen you can manage all of your home screens. To delete a home screen completely, tap on the trash icon at the upper right. To change the order of your home screens, tap on a home screen, hold your finger for a second, and then drag the home screen to a new position (left or right). To create a new home screen, swipe your finger to the left until you reach a blank home screen with a plus symbol inside of it. Tap on this plus symbol to create a new home screen.

How to Add Widgets to your Home Screens

1. Enter the home screen editing tool by tapping and holding down on a home screen until your home screen breaks away.
2. Tap Widgets at the bottom.
3. Here all of your available widgets will be shown (Figure 16.3). You can browse between widgets by swiping left and right. Widgets are sorted by the app the widget is derived from, so if there is a number in parenthesis under the app name, that means that there are multiple widgets to choose from. If a size is shown under the app name (i.e. 3x2), that indicates the minimum size of the widget (i.e. 3 tiles by 2 tiles).

Chapter 16 | The Home Screen

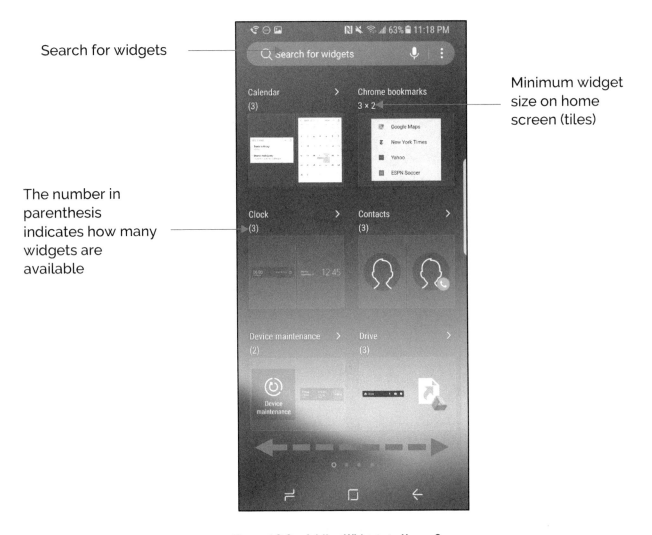

Search for widgets

Minimum widget size on home screen (tiles)

The number in parenthesis indicates how many widgets are available

Figure 16.3 – Adding Widgets to Home Screens

4. Tap and hold on a widget to add it to the current home screen. In this example, I will be adding the Alarm widget found in the Clock selection (Figure 16.4).

Chapter 16 | The Home Screen

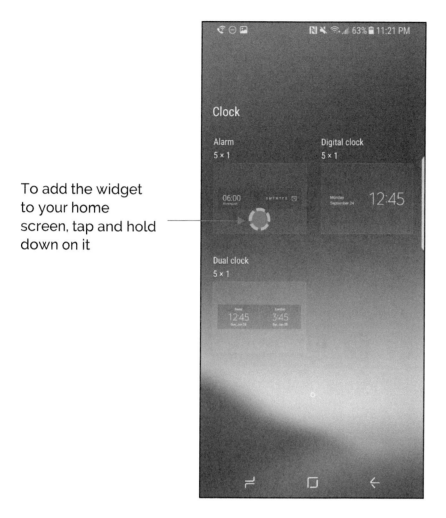

To add the widget to your home screen, tap and hold down on it

Figure 16.4 – Adding Clock -> Alarm Widget to Home Screen

5. Now, while holding your finger down on the widget, you can drag it to a location on your home screen. You will notice markers on your home screen indicating tile size (Figure 16.5).

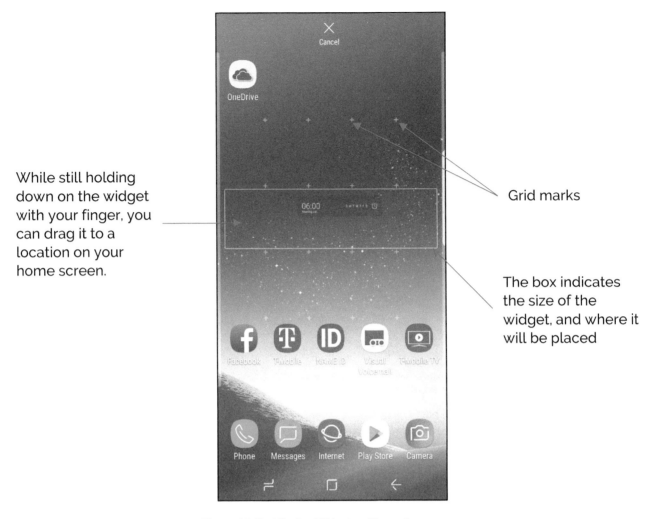

Figure 16.5 – Placing Widget on Home Screen

6. Once you have found an appropriate location, let go of the widget, and it will be placed.
7. If you are allowed to adjust the size of the widget, a turquoise box will appear around the widget with small circles. You can tap and drag the dots to adjust the size of the widget (Figures 16.6 & 16.7).

Chapter 16 | The Home Screen

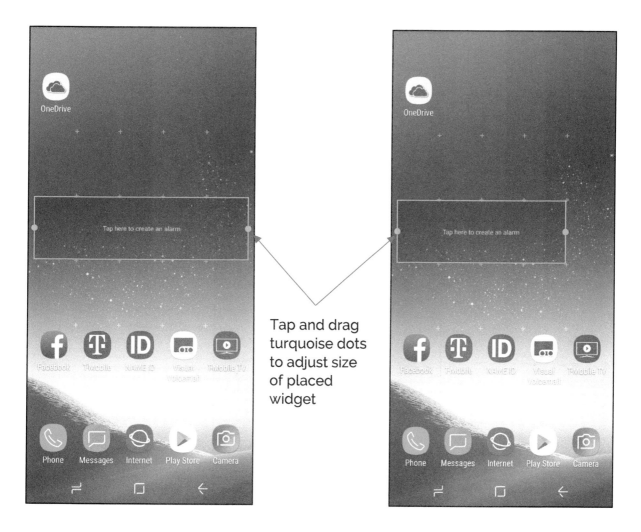

Figure 16.6 – Adjusting Size of Placed Widget (1)

Figure 16.7 – Adjusting Size of Placed Widget (2)

8. Once you are satisfied with the size of the widget on your screen, tap on an area of your home screen not inside the widget and the widget will be set and placed (Figure 16.8).

Chapter 16 | The Home Screen

Figure 16.8 – Widget Placed on Home Screen

9. Now you are free to use the widget from your home screen by tapping on it.

Your choices of widgets that you can add to your home screen will be limited at first. As you download more apps, more widgets will become available.

How to Adjust the Size of a Widget once it has already been placed

1. Tap and hold down on the widget on your home screen until the turquoise box appears (Figure 16.9).
2. Now you can adjust the size by tapping and dragging on the turquoise circles.
3. You can move the entire widget by tapping on the center of it, holding down, and dragging the widget to a new place on your home screen. You can move it to a different home screen by dragging it to the edge of the screen.
4. Tap on an area of your home screen off the widget when you are finished.

How to Delete a Widget from a Home Screen

1. Tap and hold on a widget until a box appears near it (Figure 16.9)

Chapter 16 | The Home Screen

2. Tap Remove

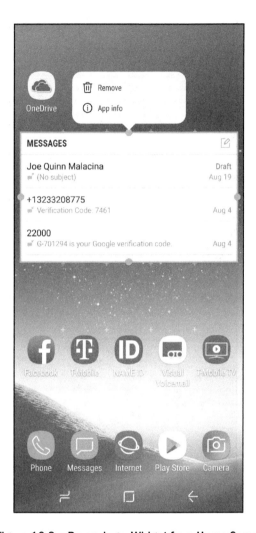

Figure 16.9 – Removing a Widget from Home Screen

How to Add Apps to your Home Screens

1. Go to the home screen you want to add an app to
2. Open the Apps page by tapping on your home screen and swiping up
3. Find the app you want to add to the home screen, and tap and hold on the app
4. Wait until your home screen appears, then drag and drop the app to the desired location (Figure 16.10)

Chapter 16 | The Home Screen

Tap and hold on an app in the Apps page until your home screen appears, now drag and drop it to desired location

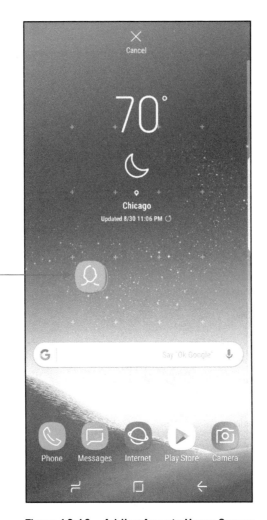

Figure 16.10 – Adding Apps to Home Screen

5. You can move any app that already appears on a home screen by tapping and holding it until a white box appears nearby, and then dragging it to the desired location (Figure 16.11)

Chapter 16 | The Home Screen

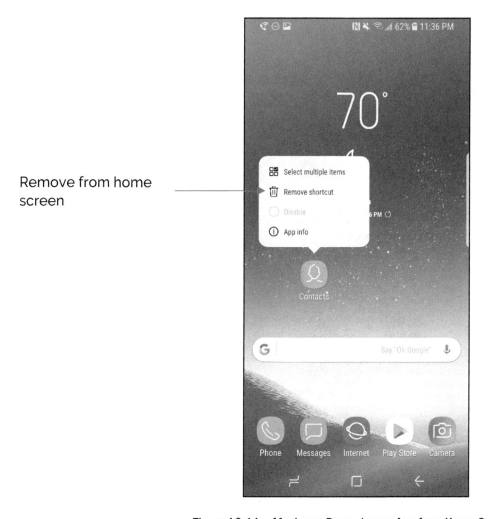

Remove from home screen

Figure 16.11 – Moving or Removing an App from Home Screen

How to Remove an App from the Home Screen

1. Tap down on an app that is on your home screen and hold until the white box appears (Figure 16.11)
2. Tap Remove shortcut

Please note, this will not delete the app from your Galaxy, it will merely remove the app from your home screen.

How to Delete an App from your Galaxy device

1. Open the Apps Page by tapping on your home screen and swiping up
2. Tap and hold on the app you want to delete until the white popup box appears
3. Tap Uninstall (Figure 16.12)

Please note, some apps cannot be deleted, such as the ones that came with your Galaxy or are required to perform certain functions. In these cases, the option to disable the app may appear, or no option of this type may appear at all.

Chapter 16 | The Home Screen

Tap and hold on an
app in the Apps Page
to bring up the menu

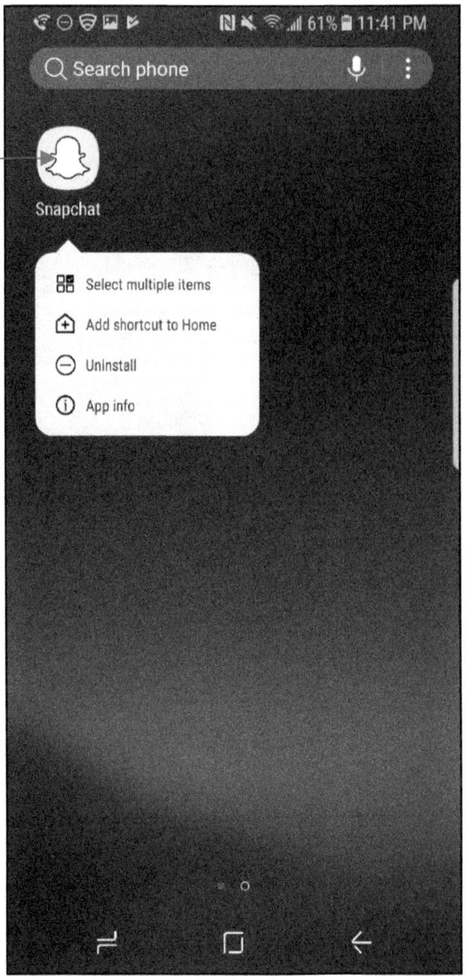

Figure 16.12 – Uninstalling an App from Galaxy

Chapter 17 – Notifications

So far we have covered all the basics of using the Galaxy S8. You now know how to navigate the Galaxy, make phone calls, exchange text messages, browse the internet, use your email, personalize your device, download apps, and more. Now we are going to get into specific features that are absolutely essential to understand, and we start with notifications.

Overview

Notifications are an integral part of your Galaxy. Every time you receive a text message, you will also receive a notification. Every time you miss a phone call, you will receive a notification. In fact, every time certain information is delivered to you from an app on your Galaxy, you will receive what is called a notification.

Notifications appear on your Galaxy in several forms. On your lock screen, you will see notifications lined up (Figure 17.1). As a reminder, to see your lock screen when your screen is off, simple press the power button. For notifications on your lock screen, you can tap on one and then unlock your device to go straight to the notification.

While using your Galaxy, you will see notifications in real-time appear at the top of your screen (Figure 17.2). When these real-time notifications appear, you can tap on it to be brought directly to the notification. For instance, if you receive a text message from someone while you are doing something on your Galaxy, a box will appear as in Figure 17.2, and this box will show the text message and who it is from. This box will appear for a brief period and while it is there, you can tap on it to be brought directly into the text conversation in the Messages app. Conversely, you can tap the bolded text options in the box to CLOSE, MARK AS READ, or REPLY straight from the notification.

Chapter 17 | Notifications

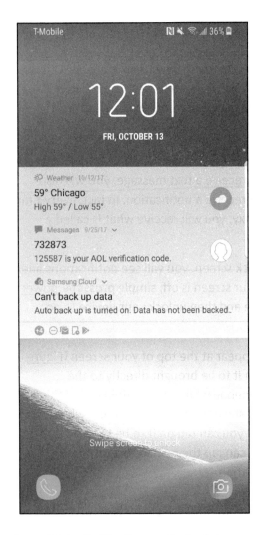

Figure 17.1 - Notifications on the Lock Screen

Figure 17.2 - Notifications when Galaxy Unlocked

Another type of notification is the icon notification. This notification is the orange circle with the number inside of it appearing at the top right of an app, such as the Messages app in Figure 17.2. This notification usually refers to a number of unacknowledged notifications you have for that specific app. For instance, the number 3 inside in the Messages app usually means I have three unread text messages. Once I open the Messages app and read those messages, the icon notification should disappear.

The Notification Bar

Sometimes referred to as the Notification Panel, the Notification Bar is where all of your notifications go. At any time you can view all of your current notifications by tapping down at the top of your screen and swiping down. This will bring up the Notification Bar (Figure 17.3). Here you can see all of the notifications that you have not acknowledged. To go to a particular notification, simply tap on it. To clear a notification, simply tap on it and swipe it fully left or right. You can clear all of your notifications by tapping on CLEAR ALL at the bottom of the Notification Bar. To return to your home screen, either swipe up or press the home button.

116

Chapter 17 | Notifications

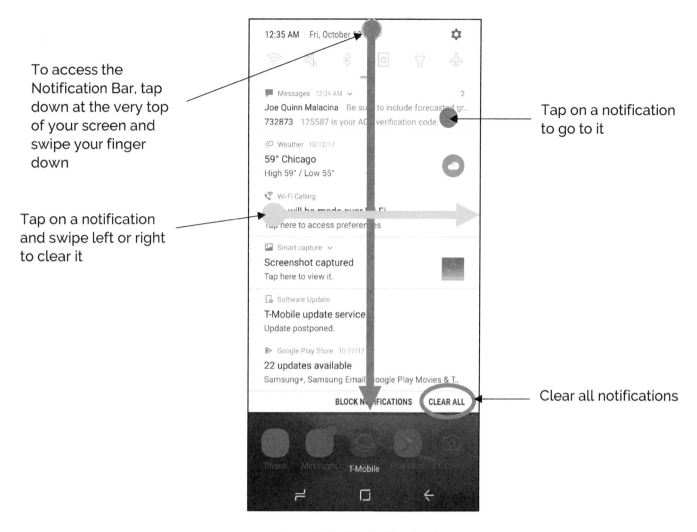

To access the Notification Bar, tap down at the very top of your screen and swipe your finger down

Tap on a notification and swipe left or right to clear it

Tap on a notification to go to it

Clear all notifications

Figure 17.3 – The Notification Bar

Examples of Notifications

As stated earlier, a notification can be from any app. For instance, a breaking news story from Fox News can be a notification if you have the Fox News app. A new Facebook like on your profile can be a notification if you have the Facebook app installed. Basically, any app can deliver notifications.

Managing Notifications

Sometimes, notifications can become a real headache and you may want to turn notifications off from a certain app. This can be accomplished easily in the Settings app.

1. Open the Settings app in the Apps page.
2. Tap Notifications.
3. Scroll and find the app you want to manage notifications for and tap on it.
4. Now you can disable notifications for this app by tapping on the blue slider icon next to Allow notifications (Figure 17.4). When this slider icon is no longer blue and tabbed to the left, notifications will be turned off for that app.

117

Chapter 17 | Notifications

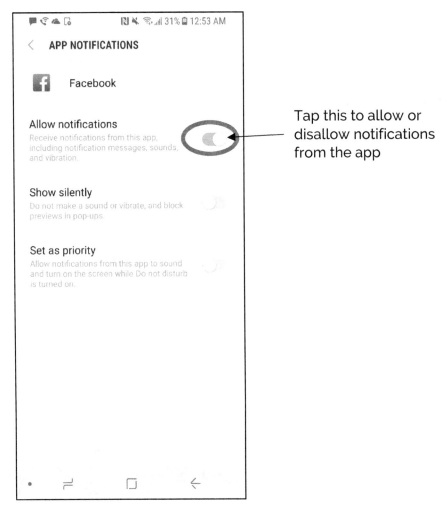

Figure 17.4 – Settings App -> Notifications -> App Name (Facebook)

Chapter 18 – The Notification Bar

We have just covered the Notification Bar in Chapter 17. We showed you how to access it, access notifications, and clear notifications. However, there is so much more you can do with the Notification Bar that it deserves its own chapter. The Notification Bar is incredibly useful, and you may find yourself using it often to perform essential tasks on your Galaxy. So let us open the Notification Bar and explore what else we can do. Remember, to open the Notification Bar tap down at the very top of your screen and swipe down.

Quick Controls

There are a number of quick controls you can access from the Notification Bar and these can be seen in Figure 18.1.

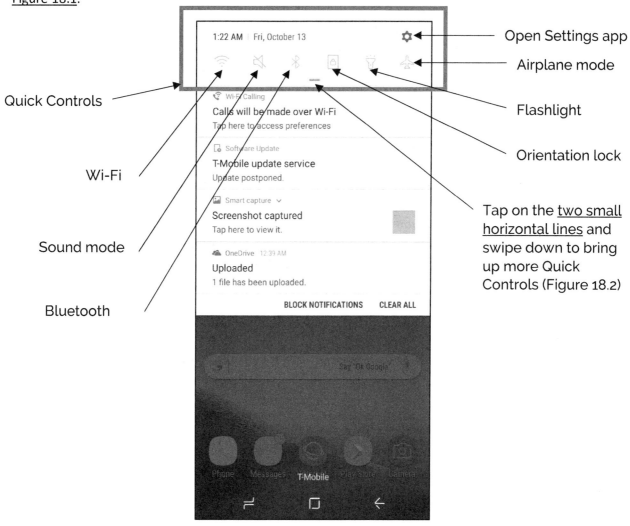

Figure 18.1 – The Notification Bar Quick Controls

At the top of the Notification Bar will be some icons, some of which may be highlighted. These are known as Quick Controls. Tapping on these icons will perform a specific function, such as turn on vibrate

119

mode, enable Bluetooth, and more. These icons are an efficient way to perform functions and tasks on your Galaxy. When an icon is highlighted, it means it is currently enabled. Here are what the Quick Control icons do as in Figure 18.1. (Note: you may see different icons on your screen but it will not matter as all of the icons will be covered in this chapter.)

- **Wi-Fi –** The Wi-Fi symbol, which is the first icon that is on the left in Figure 18.1, enables or disables Wi-Fi. When the symbol is blue, Wi-Fi is on. Use this icon when you need to quickly disable Wi-Fi and again to enable it.
- **Sound Mode –** The speaker icon is used to control the sound mode of your Galaxy. There are three different modes you can quickly switch to using this icon. Each time you tap on the icon the mode will change along with the icon itself. You can see which mode you are in by looking at the icon.
    - **All Sounds On:** The speaker icon will show sound waves emanating from it and will make a sound when enabled.
    - **Vibrate Mode:** The speaker icon will have a diagonal line crossing through it and a vibrate symbol emanating from speaker. Your Galaxy will also vibrate when this is enabled. No sounds will play when this mode is enabled and your Galaxy will vibrate in its stead.
    - **No Sounds Mode:** The speaker icon will have a diagonal line crossing though it and no symbols will be emanating from the icon. This turns off all sounds on your Galaxy and turns off vibration as well.
- **Bluetooth –** Enables or disables Bluetooth. To connect to a Bluetooth device, go to Settings -> Connections -> Bluetooth.
- **Orientation Lock –** Allows you to lock the orientation of your Galaxy in portrait mode or turn to Auto rotate. When in Portrait Mode (a rectangle with a lock inside of it icon), your Galaxy will not re-orientate its screen when you turn your device to its horizontal. Likewise, when Auto rotate is enabled (two circular arrows icon), your Galaxy will automatically re-orientate its display when you turn your device.
- **Flashlight –** Tapping this icon turns on the flashlight, which is the Galaxy's flash. This flashlight is very bright in small areas.
- **Airplane Mode –** When Airplane mode is on, your Galaxy will disable Wi-Fi, cellular data, Bluetooth, and cellular connection.
- **Settings App –** The gear icon open the Settings app on the Galaxy. Using this is a very quick way to open Settings.

There are even more Quick Controls you can access by tapping on the two small horizontal lines and dragging down (See Figure 18.1). This will show more Quick Controls with their indicated functions (Figure 18.2).

Chapter 18 | The Notification Bar

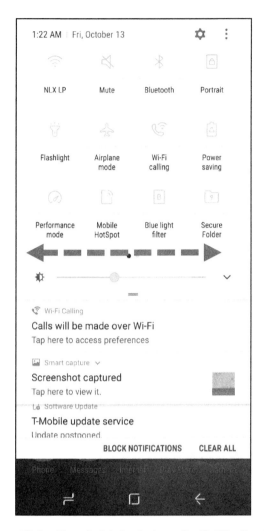

Figure 18.2 – More Quick Controls on the Notification Bar

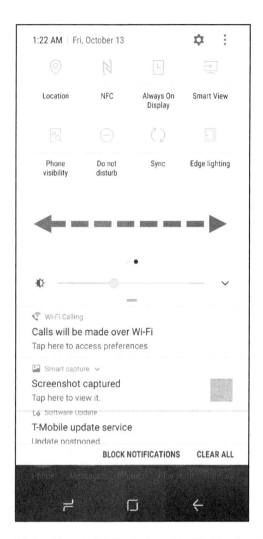

Figure 18.3 – More Quick Controls on the Notification Bar (2)

- **Wi-Fi calling** – Enables Wi-Fi calling which uses your wireless network connection to make and receive phone calls. This must be allowed and activated by your cellular provider.
- **Power saving** – Enables power saving mode which uses less battery life. This is an excellent function to use when your battery is low and it is going to be a while before you can charge it.
- **Performance mode** – The opposite of power saving mode. This mode gives you the best performance on your device but uses the most battery and processing power.
- **Mobile HotSpot** – Enables your Galaxy's Mobile HotSpot feature. This broadcasts your Galaxy's cellular data connection as a Wi-Fi network and allows other devices to connect to it. In other words, it is a way to share your internet connection with other devices. This must also be allowed by your cellular provider and may incur additional charges.
- **Blue light filter** – Adjusts the display of your screen for bedtime viewing. The desired effect is that the new display will not strain your eyes as much.
- **Secure Folder** – Opens the Secure Folder on your Galaxy which is a way to protect certain data. This is a more advanced function.

121

- **Brightness Slider –** The slider underneath the icons. Use this to manually adjust the brightness of your screen.

You can access even more Quick Controls from Figure 18.2 by swiping to the left (Figure 18.3).

- **Location –** Enables or disables Location tracking on your Galaxy. Leave this on if you want to be able use GPS and relatable functions.
- **NFC –** Enables or disables Near Field Communication, which is a way to transfer data such as photos to another device that is near you.
- **Always On Display –** Enables or disables the Always On Display, which is the black screen that shows the clock when your screen is off.
- **Smart View –** Broadcast your screen to another device such as a Chromecast.
- **Phone Visibility –** When enabled, your Galaxy will appear when nearby devices search for it using tools like Bluetooth or NFC.
- **Do not disturb –** This is one of my favorite functions. When Do not disturb is enabled, your Galaxy will play no sounds, will not vibrate, and will not light up for notifications *when your screen is off*. However, when you are actively using your Galaxy, you will receive notifications and sounds (if enabled) when Do not disturb mode is on.
- **Sync –** When enabled, your Galaxy will periodically sync with your accounts, such as your Samsung account, Google account, and mail accounts. When disabled, you will have to sync these accounts manually.
- **Edge lighting –** This is a neat feature on the Galaxy S8. Edge lighting is a notification system where the edges of your Galaxy's screen will light up whenever you receive a notification. It is a fancy way to receive a notification and is worth enabling to see how you like it.

Chapter 19 – Bixby & Google Assistant

Bixby and Google Assistant are both intelligent voice assistants that you can use on your Galaxy S8. With these services, you can quickly perform tasks using just your voice. Both services are similar, but they do have some differences. The biggest difference is Bixby is a Samsung service and Google Assistant is a Google (Android) service. As for which one you should use, that is completely up to you. I would recommend giving both a try in the beginning, and then sticking with the one you like best after some time. These services are an excellent way to perform quick tasks without having to look or use your screen.

Setting up Bixby

If you refer back to Figure 2.1 in Chapter 2, there is a button on the S8 we have not used yet, and that is the Bixby button. You can press this button, which is located on the left side of your device underneath the volume buttons, to open Bixby and be brought through the setup process. Simply follow the instructions on your screen to get through the process. You will be asked to select your language and accept terms and conditions. Once you have gotten past the first two steps, your Galaxy S8 will teach you how to use Bixby. Following this, you will have to train your device to recognize your voice. The instructions on your screen are very straight-forward, and when you are finally finished, you will be able to use Bixby at any time.

Using Bixby

Once Bixby is set up, using it is simple. Whenever you want to speak to Bixby, press and hold the Bixby button on the side of your Galaxy and speak your command. Release the Bixby button once you are finished speaking. Bixby will listen and try to execute what you commanded.

There are countless things you can say to Bixby that she will perform. Try out anything you would like. Here are some examples:

- Make Phone Calls – "Call *contact name*"
- Lookup contact information – "Lookup *contact name*"
- Lookup internet information – "What time do the New England Patriots play tonight?"
- Set an alarm clock – "Set an alarm for 6:30 AM"
- Do math – "What is two plus two?"
- Visit Websites – "Bring me to InfinityGuides.com"

Bixby will listen to every command you tell her while the Bixby button is held down. She can respond by opening an app, speaking back a response, or opening a screen with the desired result.

You can also access the Hello Bixby screen by pressing the Bixby button and quickly releasing (Figure 19.1). This screen will just show you content Bixby thinks is relevant to you.

Chapter 19 | Bixby & Google Assistant

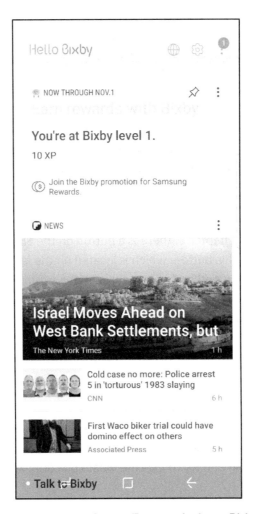

Figure 19.1 – Hello Bixby Screen (Press and release Bixby button)

Setting up Google Assistant

To set up Google Assistant, press and hold on the home button until a new window appears (Figure 19.2). This is Google Assistant, which works similarly to Bixby. You may be guided through a setup process the same way you were for the Bixby setup.

Chapter 19 | Bixby & Google Assistant

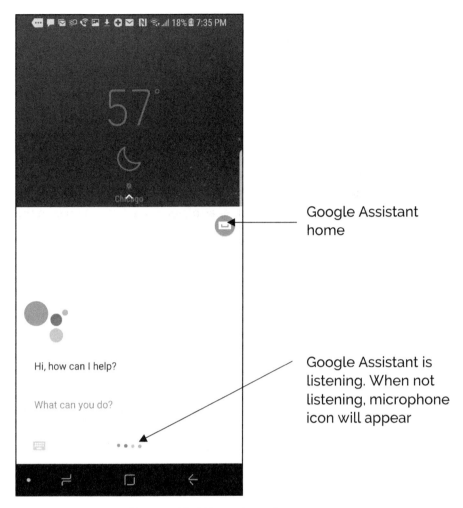

Figure 19.2 - Google Assistant (Press and hold home button)

Using Google Assistant

To use Google Assistant at any time, simply press and hold the home button until the Google Assistant window opens. Then, while still holding on the home button, speak your command to your Galaxy, releasing the home button when you are done speaking. Google Assistant will listen and try to execute your command. You can give another command by tapping on the microphone icon at the bottom of the Google Assistant screen or by pressing and holding the home button again. You can command Google Assistant to do just about anything, including the same examples listed in the Using Bixby section of this chapter. One notable aspect of Google Assistant is that it is linked with your Google account, therefore you can use Google Assistant to explore features of your Google account that Bixby may not be able to do. For instance, you can ask Google Assistant: "Do I have any emails from John Doe?" Google Assistant will search your Google emails automatically and bring up any emails from John Doe. If you were to do this command using Bixby, Bixby would open your Mail app and tell you to type in your search terms in the search box.

Here are some more examples of what you can say to Google Assistant:

- Open an app – "Open the Play Store"
- View text messages – "Show me my messages"
- Send email – "Send a new email to *contact name*"
- Check weather – "What is the weather in Chicago?"

Bixby vs. Google Assistant

As stated earlier, you can use both services to accomplish what you need. Bixby is more integrated with native apps on your device. So you may want to use Bixby when you are looking to perform specific app functions such as changing settings. On the other hand, Google Assistant is great at performing tasks related to Google apps such as Gmail. Google Assistant is also excellent for searching for information on the internet. There is no clear winner as of yet in the battle between these two services, so stick with who you like.

Chapter 20 – Native Apps

Let us explore some native apps that come preloaded on the Galaxy S8. Many of these are very useful and you may find yourself using them often. All the apps covered in this Chapter can be found on the Apps page of your Galaxy. If any of these apps do not appear, you can download them in the Play Store.

Different Apps with Different Carriers

Your Galaxy S8 comes preloaded with many apps. We have gone over most of them throughout this text, including the Phone app, Messages app, Settings app, Play Store app, Chrome app, Email app, Camera app, and Gallery app. However, your Galaxy S8 may have apps that are very similar to the ones that we have covered. This is because many cellular carriers load their own apps onto the Galaxy S8. So if you purchased your Galaxy S8 from T-Mobile, your device may have some T-Mobile based apps on it. This can include apps for texting, voicemail, navigation, email, and much more. The same can be said for Verizon, ATT, or any other cellular carrier. These apps can be useful but your Galaxy already has apps that can perform all the basic functions needed. Therefore, in this text we will only cover apps that are native to the Galaxy S8, not apps that cellular providers preload on the device. We will also cover some Google apps that are essential on any Galaxy.

Play Music

The Play Music app is where you can download and listen to music. There are three different ways you can acquire music on your Galaxy. The first way is to purchase music right through the Play Music app. This app works much the same way as the Play Store. You can search for music using the search bar at the top or browse through categories and charts using the menu (Figure 20.1). When you try to buy a song or album, you will be redirected to the Play Store where you can purchase it.

Chapter 20 | Native Apps

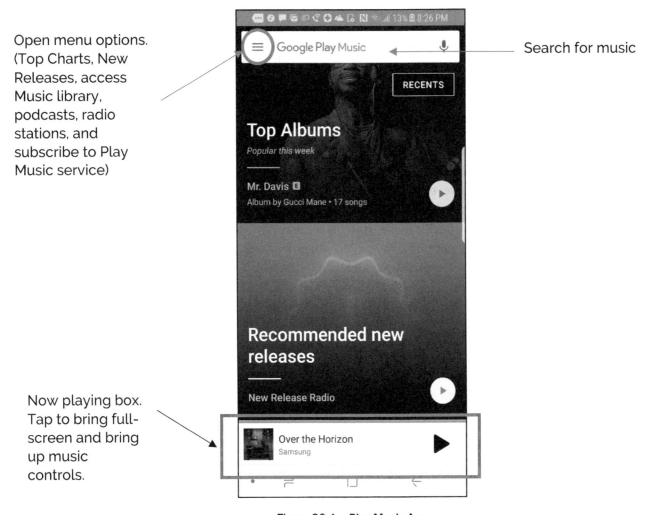

Open menu options. (Top Charts, New Releases, access Music library, podcasts, radio stations, and subscribe to Play Music service)

Search for music

Now playing box. Tap to bring full-screen and bring up music controls.

Figure 20.1 – Play Music App

The second, and likely best way to listen to music is to subscribe to Play Music. When you subscribe to Play Music, you can you listen to the entire Play Music library at any time from any device. Subscribing costs $9.99 per month, and sometimes there is a free 3-month trial available. To subscribe to the Play Music service, tap on the menu icon, then tap SUBSCRIBE NOW.

The third, and final way to get music on your Galaxy S8 is to transfer music from your computer. To do this, simple plug your Galaxy S8 into your computer using the Samsung USB cable. Then use Finder (Mac) or Windows Explorer (Windows) to navigate to your Galaxy device, it will usually be called SAMSUNG-SM-XXXXX and can be found in the same location as C:/ (My PC). Then navigate to Phone -> Music. From here, you can drag and drop any music files on your computer into this Music folder, and it will automatically be placed into your Play Music library.

To Listen to your Music Library

Once you have created your music library, whether by buying songs, subscribing to Play Music, or transferring songs from your computer, you can view and listen to all of it in your Music library. To access the music library:

Chapter 20 | Native Apps

1. Open the Play Music app
2. Tap the menu icon (three horizontal lines)
3. Tap Music library

Here inside the Music library you can browse through your music using the menus at the top, and simply tap on a song to play it (Figure 20.2). You can tap on the song that is playing at the bottom to bring up the music controls.

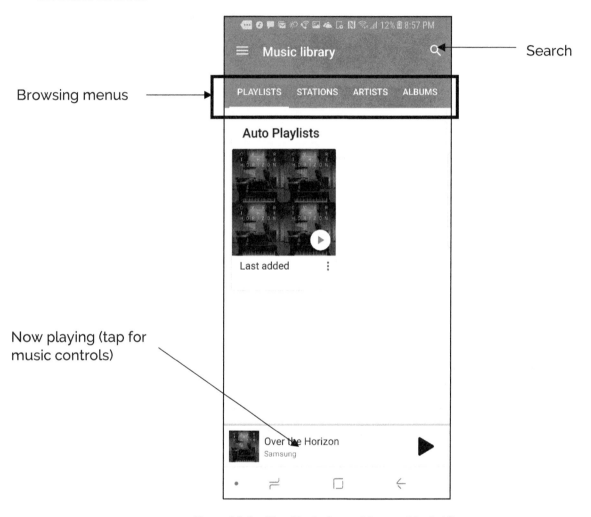

Figure 20.2 – Play Music App -> Menu -> Music Library

Google Maps

The Google Maps app is one of the best apps for navigation, searching for places, and checking traffic. In fact, Google Maps was rated the #1 navigation app by Infinity Guides for the year 2017. You can use Google Maps for a variety of purposes (Figure 20.3). One of its most popular features, is its turn-by-turn navigation. In other words, it is an excellent GPS for directions. You can also use Google Maps to search for nearby places such as restaurants, gyms, schools, etc. You can also read user reviews of places right from within the app. In addition, you can turn on Satellite view, view current traffic, and get driving times.

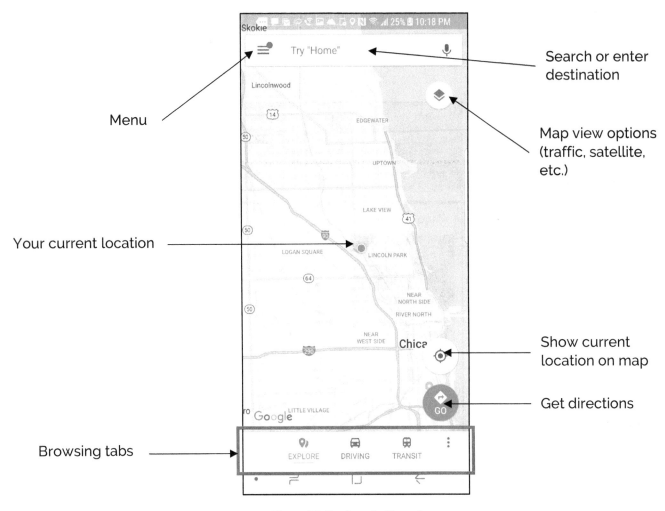

Figure 20.3 – Google Maps App

How to get Turn-by-turn Directions

1. Tap into the search bar at the top of Google Maps.
2. Type in your destination (address or name of place).
3. Google Maps will offer suggestions as you are typing. You can tap on the suggestion to autofill or tap on the search icon on your keyboard to execute search.
4. Tap the DIRECTIONS box.
5. Tap START to begin turn-by-turn directions (Figure 20.4).
6. Directions will appear on your screen and update as you are moving. Upcoming turns will play from your Galaxy's speaker if you have sound enabled.

Chapter 20 | Native Apps

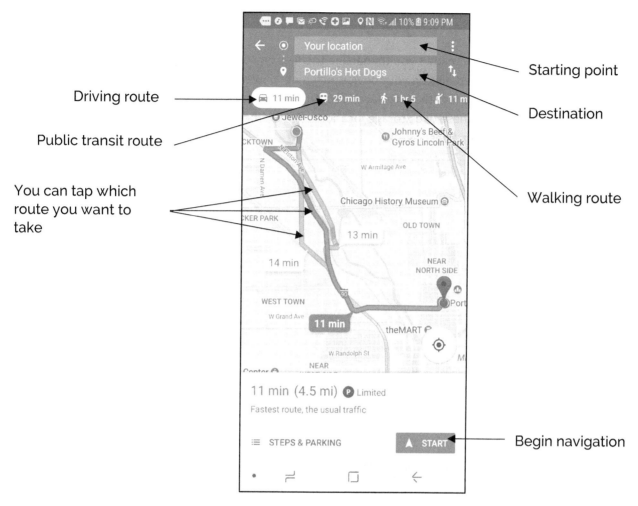

Figure 20.4 – GPS Navigation

How to Search for Nearby Places

1. Tap into the search bar
2. Type in a search term, such as "restaurants". Use autofill or tap the search icon to execute the search.
3. Google Maps will return a list of places along with a map view. You can tap on a list entry to see more information including operating hours, reviews, and directions.

At any time, you can zoom in and out of the map using two fingers, the same way you can zoom while browsing the internet (see Chapter 10).

Internet

The Internet app is Samsung's web browser, and an alternative to Google Chrome which we covered in Chapter 10. The app works in much the same way as Chrome, with some minor differences in navigation. This app is mentioned just so you are aware that there are two different internet browsers available on the Galaxy S8.

My Files

The My Files app is a nifty tool that allows you to manage the files on your Galaxy (Figure 20.5). Within this app, you can view all the saved files on your device including: photos, videos, PDFs, downloads, songs, and more. You can also see files that are saved on your device and files that are saved on your SD card, and transfer files between these two storage locations. (Note: The My Files app may be located inside the Samsung folder on your Apps page.)

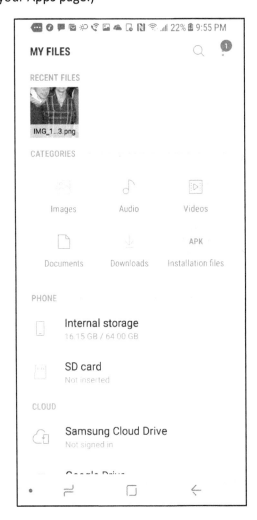

Figure 20.5 - My Files App

How to Delete Files

1. Find the file you want to delete by browsing either by category or storage device.
2. When the file is found, tap and hold on it to select it with a checkmark.
3. Tap DELETE at the upper right.

How to Move Files from Phone to SD Card or Vice Versa

1. Find the file(s) you want to delete by browsing either by category or storage device.

2. When found, tap and hold on the file to select it with a checkmark. You can do this with folders as well or select multiple files.
3. Tap the 3 vertical dots icon at the upper right.
4. Tap Move.
5. Tap the appropriate storage device (Internal storage, SD card, or cloud).
6. Choose the destination within the storage device if applicable.
7. Tap DONE.

Clock

The Clock app is a straight-forward application that you can find in the Apps page on your device. As a reminder, to access the Apps page touch down on your home screen and swipe up. Here inside the Clock app, you can use the tabs at the top to browse different features (Figure 20.6).

- **ALARM** – Create and set alarm clocks.
- **WORLD CLOCK** – View time around the world
- **STOPWATCH** – Use a stopwatch
- **TIMER** – Set a timer

How to Set an Alarm Clock

1. Open the Clock app.
2. Tap the ALARM tab at the top.
3. Tap the plus symbol in the cyan circle.
4. Fill out the alarm as necessary including time, snooze, and repeat.
5. Tap SAVE at upper right when done.
6. You can quickly enable or disable previous alarms by tapping on the slider tab to the right of an alarm.

Chapter 20 | Native Apps

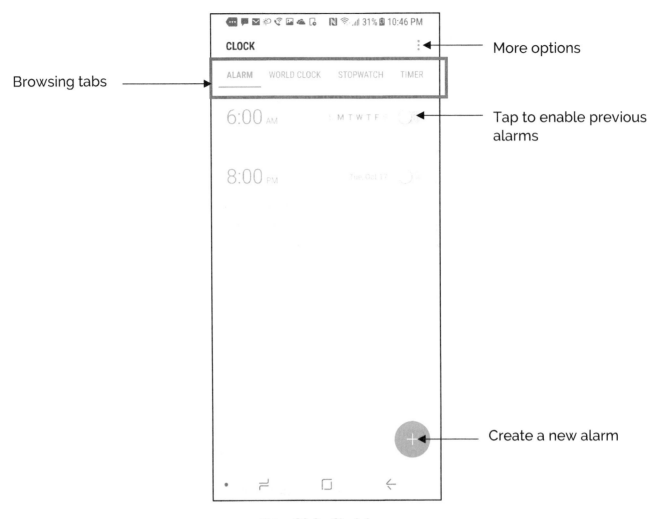

Figure 20.6 – Clock App

Calendar

The Calendar app is a simple app that allows you to create events and reminders (Figure 20.7). These events will integrate with the rest of your Galaxy to help you stay organized. For instance, when you create an event, that event will appear in your Notification Bar and lock screen as it approaches.

How to Create a Calendar Event

1. Open the Calendar app
2. Tap the plus icon at the lower right
3. Fill out the form as necessary
4. Tap SAVE at the upper right when done
5. Your event will now be saved to your Calendar

When viewing your Calendar, events will be shown as highlighted text under a date. To view all events on a date, tap on the date. To then view information about the event, tap on the event name inside the

Chapter 20 | Native Apps

date box. You can also quickly add an event to a date by tapping on the date, and then tapping on the plus icon.

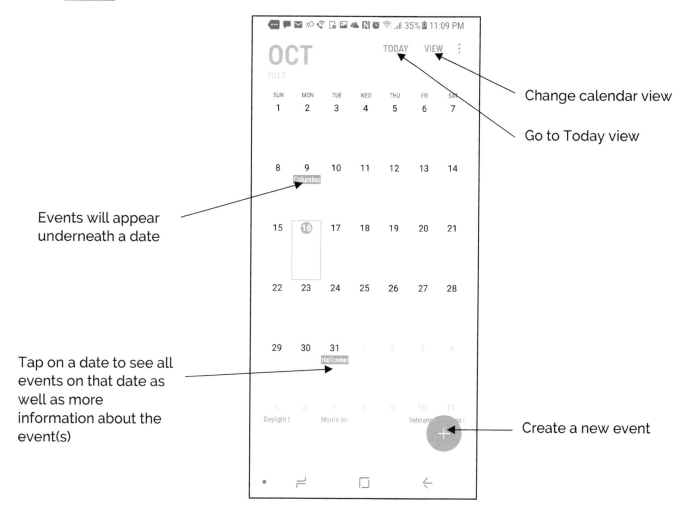

Change calendar view

Go to Today view

Events will appear underneath a date

Tap on a date to see all events on that date as well as more information about the event(s)

Create a new event

Figure 20.7 – Calendar App

135

Chapter 21 – Tips & Tricks

Congratulations! You have made it through the majority of this book, and you should now have a solid and fundamental understanding of using your Galaxy S8. There are no more basics left to teach you, so I will leave you with a few tips and tricks that you may find helpful when using your Galaxy.

Backing up your Device

This is an extremely important tip. You should always have a backup of your Galaxy's data available in case anything happens. If you destroy or lose your Galaxy, a backup will allow you to recover all of your data including contacts, photos, and videos. There are two types of backups you can enable.

1. **Samsung Account Backup**

You can use your Samsung account to backup your device, and all of your data on it. This is a cloud backup service which comes with a set amount of free storage which should be enough for a user with less than 200 photos on their Galaxy. The Samsung Account Backup will back up everything including: your contacts, photos, videos, apps, wallpapers, layouts, files, call log, text messages, music, and settings. Here is how you can turn Samsung Account Backup on:

1. Open the Settings app.
2. Tap Cloud and accounts.
3. Tap Backup and restore.
4. Under SAMSUNG ACCOUNT, tap Backup settings.
5. Make sure AUTO BACK UP is enabled, and all aspects you want to be backed up are enabled.

Since Samsung Account Backup is automatic, once you have enabled it you will not need to worry about it again. Your Galaxy S8 will backup automatically every single day and your data will be recoverable if anything happens. Please note, Samsung Account Backup will not back up your contacts if they are saved to your Google account, in which case you should have Google Backup enabled as well.

2. **Google Backup**

With Google Backup, your data will be backed up to your Google account. This includes phone log, app settings, contacts, text messages, photos, videos, and more. This backup is also automatic, so once you enable it you will not need to worry about it anymore and your data will be recoverable. You also get a set amount of free storage available with your Google account. Here is how to enable Google Backup:

1. Open the Settings app.
2. Tap Cloud and accounts.
3. Tap Backup and restore.
4. Under GOOGLE ACCOUNT, tap on Back up my data.
5. Make sure it is ON. If not, tap on the tab to enable it.
6. Please note you must be signed in with your Google account on your Galaxy in order to use Google Backup.

Now your data will automatically backup to your Google account. I HIGHLY recommend using both Samsung Account Backup and Google Backup.

Taking a Screenshot

A screenshot is a picture of exactly what your screen looks like, and can be very useful. For instance, say you received a text message with a grocery list you need to fulfill, and you want to share that list with someone else. One way to accomplish this is to take a screenshot of this text message and send that screenshot to a friend.

There are two different ways to take a screenshot:

1. While viewing a screen you want to capture, press and hold the power button and the volume down button until your screen animates.
2. While viewing a screen you want to capture, place the side of your hand on one end of your screen (left or right side), then drag the side of your hand all the way across your screen to the left or right. Your screen will animate if the screenshot was successfully taken.

Screenshots will save as photos in Gallery app. You can share or edit the photo the same way you can any other photo.

Background Apps

We already know that you can use the recent apps button to quickly view apps you have recently used. We can scroll through these apps and then tap on one to quickly go to it. However, for optimal performance of your Galaxy, you should completely close down apps that you are no longer using. To do this:

1. Press the recent apps button.
2. Tap on an app, and swipe it left or right off the screen to completely close it down.
3. Do this for each app that is running until no more app windows appear.
4. Alternatively, you can tap on CLOSE ALL to close all apps at once.

Chapter 21 | Tips & Tricks

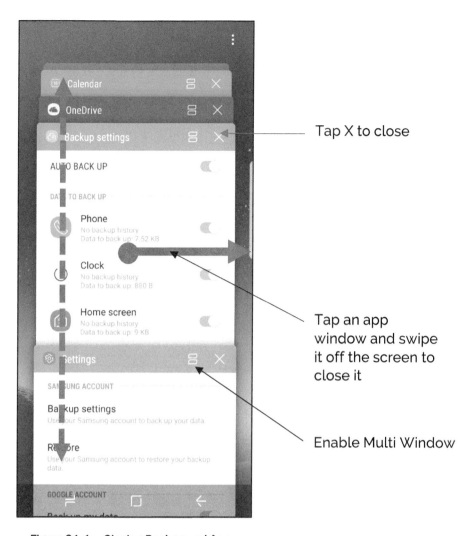

Figure 21.1 – Closing Background Apps

Multi Window

Multi Window allows you to use two apps at the same time. Here is how to do it:

1. Open an app you want to use.
2. Press the home button to go back to your home screen.
3. Open another app you want to use simultaneously with the first.
4. Press the home button to return to your home screen.
5. Press the recent apps button.
6. Tap on the Multi Window icon on the app you want to appear on the top half of your screen (see Figure 21.1).
7. Your screen will split, showing the app you just Multi Windowed on the top, and your recent apps on the bottom (Figure 21.2).
8. On the bottom half of your screen, tap on the app you want to fill this space. You can tap on MORE APPS if you want to browse other apps.

9. You are now using Multi Window. To adjust the size of the windows, tap and drag the blue line up or down.
10. To leave Multi Window, drag the blue line all the way down to the bottom of your screen.

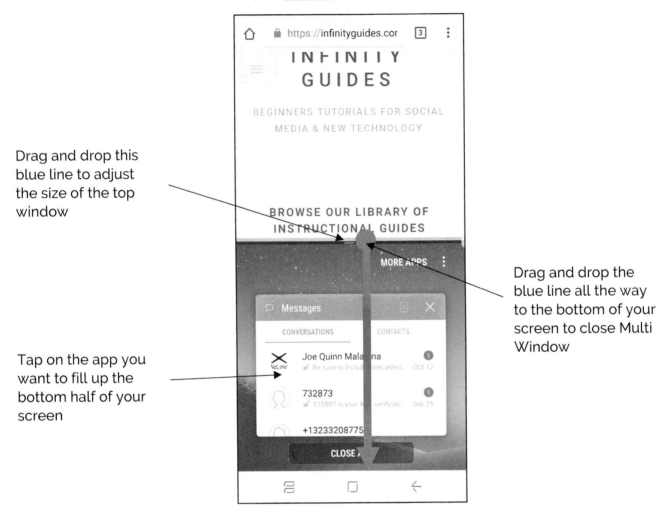

Figure 21.2 – Multi Window

Restart your Galaxy

It is best practice to restart your Galaxy every now and then, the same way you would restart a computer to keep it fresh. I recommend restarting your Galaxy if it is running slow or not working properly. Doing so can fix the problem. As a rule of thumb, I recommend restarting your Galaxy at least twice a month. To do so, simply turn the Galaxy off using the power button. Wait 60 seconds after the phone is off, then turn it back on.

Copy and Paste

To copy and paste text, find the text you want to copy and tap and hold on it. Now tap and drag the blue markers left and right to select the exact portion of text you want to copy. Now tap on COPY (Figure 21.3). You can paste the text by tapping and holding in a text area, and then tapping Paste.

Chapter 21 | Tips & Tricks

Figure 21.3 – Copy and Paste Text (Chrome app)

Status Bar

The Status Bar on your Galaxy is simply the top bar you can see on your home screen (Figure 21.4). The icons inside the Status Bar at the upper left correspond to current notifications you have. When you open the Notification Bar and clear the corresponding notification, the icon in the Status Bar will disappear. Also from the Status Bar, you can see your battery life, cell signal strength, and other Galaxy metrics.

Chapter 21 | Tips & Tricks

These icons correspond to current notifications

These icons correspond to active phone modes such as Bluetooth status and sound mode. You can also see general Galaxy metrics such as battery life and signal strength.

Figure 21.4 – The Status Bar

Updating your Galaxy

Your Galaxy will need software and app updates every so often; software updates less so. Usually, your apps will update automatically and you will be notified in your Notification Bar. Other times, depending upon your settings, you will be notified through the Notifications Bar when there are a significant number of app updates available. You can tap on this notification to update all these apps.

For software updates, otherwise known as Android operating system updates, you will receive an alert notification when one becomes available for your device. This notification will be persistent, and it is recommended that you update your operating system whenever it is available. You can also check for software updates manually in Settings -> Software update. For news and information on new software updates, I would recommend checking out www.galaxyvideoguides.com. They usually have a new video guide available for new versions of Android within a week of release.

142

Chapter 22 – More Resources

This guide has covered all the beginner aspects of the Galaxy S8. We have also covered many intermediate and advanced aspects, but there is still plenty more you can learn if you want. Most notably, it can be beneficial to learn about popular third-party apps on your Galaxy such as Facebook, Twitter, Snapchat, and Instagram.

- **Facebook Guide** – www.infinityguides.com/facebook.html
- **Twitter Guide** – www.infinityguides.com/twitter.html
- **Other Social Media Guides** – www.infinityguides.com
- **Video Tutorials on Galaxy Smartphones and Tablets** – www.galaxyvideoguides.com

Conclusion

Thank you for taking the time to read this text. It is my hope that you now feel confident about using your Galaxy S8, and I am confident that if you took the time to read this entire book, then you will have no problem using every aspect of your Galaxy with ease. Continue to use this book as a reference when you need it. The table of contents can quickly lead you to your answer, and the appendices can be especially helpful.

I welcome your thoughts and feedback on this book; please come visit my Facebook or Twitter page online at www.facebook.com/joemal or www.twitter.com/joemalacina (@JoeMalacina). I am often online answering questions from people who have read this text, and helping people with their smartphone issues. As a last piece of advice, please remember your lock screen security, and please set up automatic backup on your device. Doing these two things can save you from a very large headache down the road.

Enjoy using your Galaxy.

Appendix A – Recommended Apps

- AirBnB – Travel
- AM 560 TheAnswer – Talk Radio
- Amazon – Shopping
- Bible – Reading
- Clash of Clans – Game
- Crossword – Game
- Facebook – Social Media
- Facebook Messenger – Social Media
- Fandango – Movies
- Flixster – Movies
- Fox News – News
- Fox Sports – Sports
- Groupon – Shopping
- iHeartRadio – Radio
- Instagram – Social Media
- Lyft – Travel
- McDonalds App – Food
- Microsoft Office – Productivity
- MLB App – Sports
- Netflix – Entertainment
- OpenTable – Dining
- Periscope – Social Media
- Priceline – Travel
- Radio.com – Radio
- Shazam – Music ID
- Shopular – Shopping
- SitOrSquat – Miscellaneous
- Skype – Video Chatting
- Snapchat – Social Media
- Speedtest – Utilities
- SpotHero – Parking
- The Weather Channel – Weather
- Twitter – Social Media
- Uber – Travel
- VLC – Media Player
- Wall Street Journal – News
- Waze – Directions & GPS
- Words with Friends – Game
- Yelp – Reviews
- Your Bank's App

Appendix B – List of Common Functions

- **Add New Email to Galaxy:** Settings -> Cloud and accounts -> Accounts -> Add account -> Email
- **Change Language:** Settings -> General management -> Language and input -> Language
- **Change Lock Screen Security:** Settings -> Lock screen and security -> Screen lock type
- **Change Ringtone:** Settings -> Sounds and vibrations -> Ringtone
- **Change Wallpaper:** Settings -> Wallpapers and themes
- **Check for Android Update:** Settings -> Software Update
- **Check which Android Version you have:** Settings -> About phone -> Android version
- **Reset Galaxy to Factory Settings (WARNING: this will delete all of your data, apps, and settings and restore the Galaxy phone to right out of the box status. DO NOT DO THIS unless your data is backed up and you know what you are about to do):** Settings -> About phone -> Scroll all the way to bottom -> RESET -> Factory data reset
- **Connect to a Bluetooth Device:** Settings -> Connections -> Bluetooth
- **Connect to a Wi-Fi Network:** Settings -> Connections -> Wi-Fi
- **Create New Text Message:** Messages app -> Square and pencil icon
- **Clear Internet History (Chrome):** Chrome app -> Options (3 vertical dots icon) -> History -> CLEAR BROWSING DATA...
- **Delete Recent Call History:** Phone app -> RECENTS tab -> Tap and hold on a call -> Select calls to be deleted -> DELETE
- **Uninstall an App:** Apps page -> Tap and hold on an app until menu appears, release -> Uninstall
- **Change Storage Location of Contacts:** Contacts app -> Options (3 vertical dots) -> Manage contacts -> Default storage location
- **Export a Copy of your Contacts:** Contacts app -> Options (3 vertical dots) -> Manage contacts -> Import/Export contacts -> EXPORT
- **Sort Contacts by Last Name:** Contacts app -> Options (3 vertical dots) -> Settings -> Sort by -> Last name